Lead Pollution
Causes and control

R. M. Harrison
Ph.D.
University of Lancaster, Lancaster
D. P. H. Laxen
M.Sc., Ph.D
University of Lancaster, Lancaster

LONDON AND NEW YORK
CHAPMAN AND HALL

First published 1981
by Chapman and Hall Ltd
11 New Fetter Lane, London EC4P 4EE

Published in the USA
by Chapman and Hall
in association with Methuen, Inc.
733 Third Avenue, New York NY 10017

Printed in Great Britain at the
University Press, Cambridge

British Library Cataloguing in Publication Data

Harrison, R. M.
Lead pollution.
1. Lead – Environmental aspects
2. Pollution
I. Title II. Laxen D.P.H.
614.7 TD196.L4 80-41770

ISBN 0-412-16360-8

Contents

Preface

At the time of writing, the topic of lead pollution is the subject of an intense and sometimes heated debate. The argument centres upon possible adverse health effects arising from exposure of children to current environmental levels of lead. Such arguments now appear little closer to resolution than they did five years ago, although the development of ever more sophisticated biochemical and epidemiological techniques may eventually provide an answer.

Over the past five to ten years, as the general public has become aware of the lead issue, pressure has been put upon governments to limit emissions of lead, and hence limit or reduce the exposure of the population to the metal. Governments and governmental agencies have responded in several ways, varying between those who prefer to take little or no action on the basis that they see no cause for concern, and those who have taken firm action after concluding that the scientific and medical evidence warrants this approach.

Any effective control strategy for lead requires knowledge of the sources of environmental exposure and an understanding of the pathways of this metal in the environment. This book aims to provide such information and to explain the methods available for limiting emissions of lead from the most important sources. To put this information in context a chapter on the routes of human exposure to lead and the health effects is included.

The authors are environmental scientists without commitment to either the pro- or anti-lead lobby. Their prime interest is in understanding the nature and control of lead as an environmental pollutant. In all sections of this book, a balanced, impartial approach is the aim, with emphasis being given to reviewing the most significant published work in the field.

1 Introduction

Lead has been an important metal in human societies over many thousands of years. The low melting point, the ease with which it can be worked and the durability of lead account for its early use as a construction material. The use of lead pipes for water supply was particularly significant during the period of the Roman Empire. Lead has also been used over the years as a glaze on pottery, in cosmetics and as a means of sweetening wine [1].

The use of lead, however, has increased dramatically since the early days of the industrial revolution (Fig. 1.1). The trend over the last decade, nevertheless, runs counter to the previous exponential rate of growth, with mined lead production stabilizing at just over 2.5 million tonnes per annum (Fig 1.2). Future trends must remain somewhat uncertain, although a decline in production and consumption is unlikely.

Although workable deposits of lead are found in over 40 countries worldwide, the major mining production is concentrated in just four countries: The USSR, the USA, Australia and Canada (*ca.* 60% of the total) (Table 1.1). It is important to note that lead ores are frequently found in combination with other recoverable metals such as copper, zinc, silver and cadmium [3].

Refined lead is produced from both primary and secondary sources. Primary lead is that produced from mined ores, whilst secondary lead results from recycled materials such as battery plates and lead pipes. Recycled lead currently accounts for 14% of the world's production of refined lead (Table 1.2, cf. Table 1.1). The consumption of lead is concentrated primarily in only eight countries (Table 1.3). Storage batteries currently account for about half of the refined lead consumption in the western world (Table 1.4), whilst the production of tetraalkyllead, a petrol additive which reduces engine knock, accounts for about 10% of consumption.

Mining, smelting and refining of lead, as well as the production and use of lead-based products give rise to release of lead into the environment. This takes the form primarily of either lead-rich aqueous effluent streams, or emission of fume and dust into the air. A large part of the lead discharged into surface waters is rapidly incorporated into suspended and bottom sediments, and most of this lead will ultimately be found in marine sediments. Of greater concern, however, is the emission of lead into the atmosphere. The finer aerosol particles

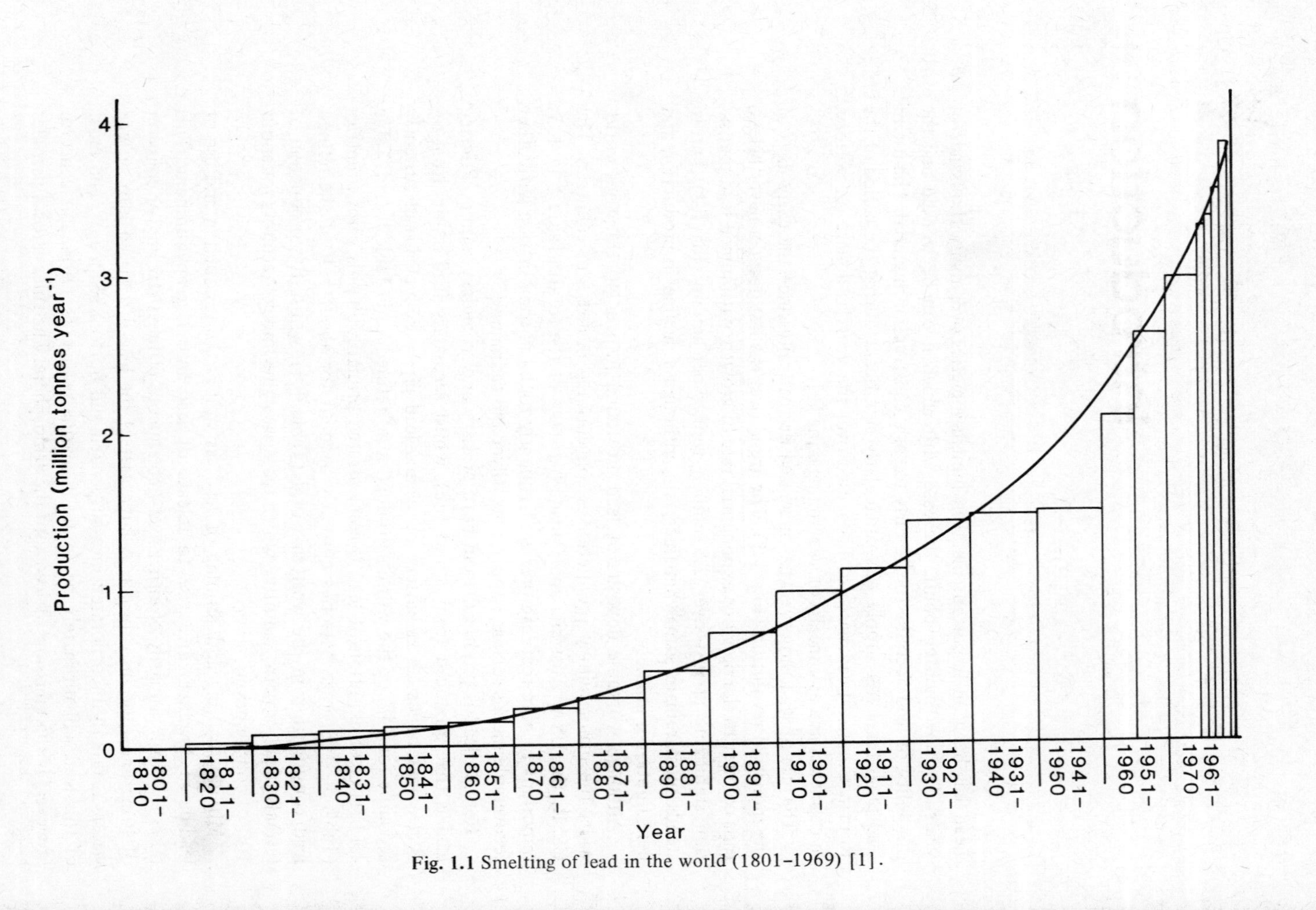

Fig. 1.1 Smelting of lead in the world (1801–1969) [1].

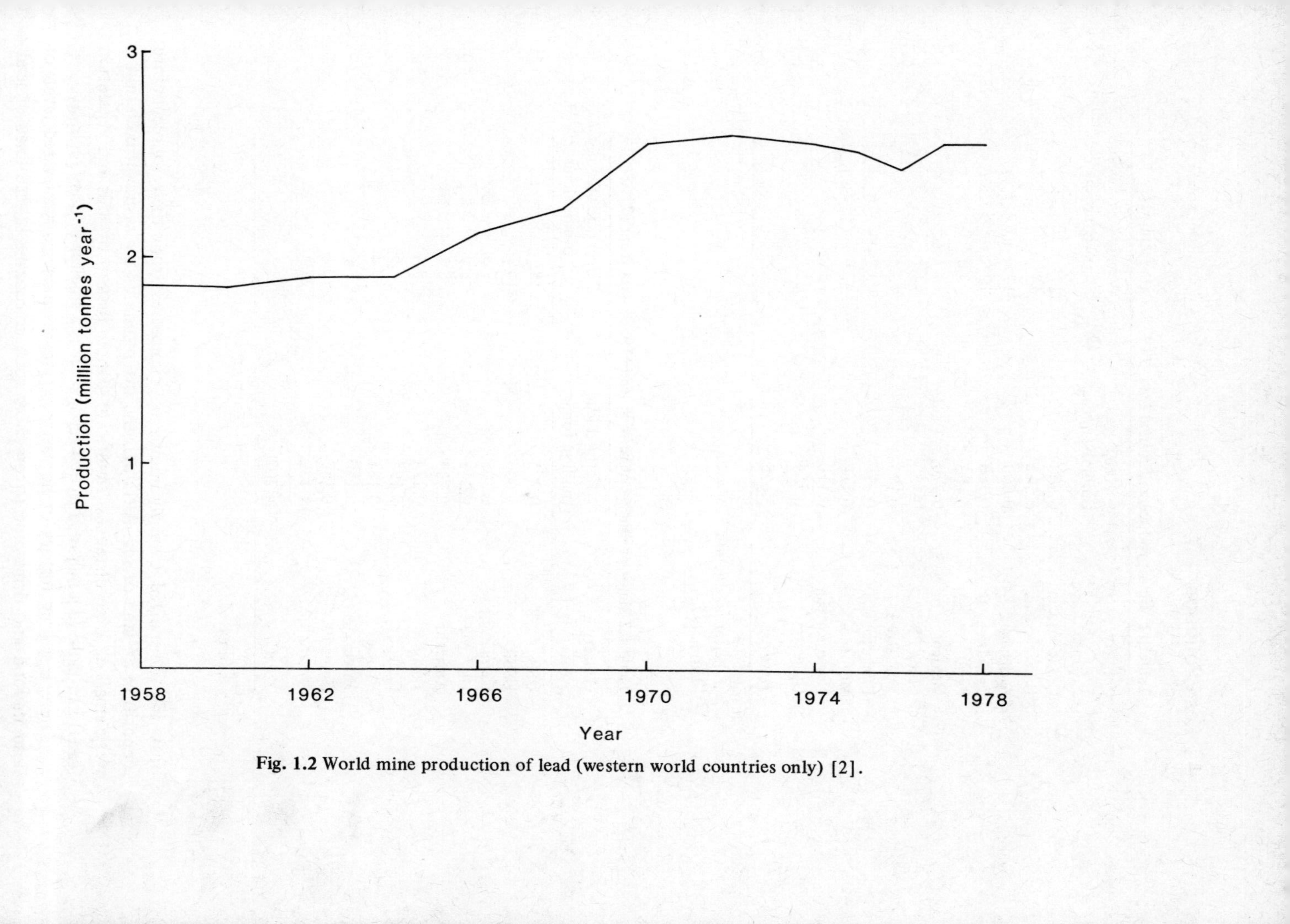

Fig. 1.2 World mine production of lead (western world countries only) [2].

Table 1.1 Major producers of mined lead* (1978) data from [2].

Country	Mined lead (million tonnes)	Percentage of world production
USSR	0.600†	17
USA	0.541	15
Australia	0.400	11
Canada	0.366	10
Peru	0.183	5.0
Mexico	0.170	4.7
China PR	0.150†	4.1
Yugoslavia	0.125	3.4
Bulgaria	0.116	3.2
Morocco	0.110	3.0
North Korea	0.110†	3.0
Others	0.754	21
Total	3.625	

*Smelter production.
†Estimated.

Table 1.2 Major producers of refined lead (1978) data from [2].

Country	Refined lead (million tonnes)	Percentage of world production
USA	0.773	18
USSR	0.600*	14
Germany FR	0.305	7.3
UK	0.247	5.9
Australia	0.239	5.7
Japan	0.228	5.4
Canada	0.194	4.6
France	0.184	4.4
Mexico	0.150	3.6
China PR	0.150*	3.6
Others	1.132	27
Total	4.202	

*Estimated.

may be transported over thousands of kilometres from their source before deposition on land or sea. Although the magnitude of the resultant pollution is very small at large distances, significant contamination of soils and vegetation can occur close to a major source of lead, such as a smelter or busy highway.

There is a long history of human exposure to abnormally elevated levels of lead in food and drink due to practices such as cooking in lead-lined or lead-glazed pots and the supply of water through lead pipes [1]. Also, some communities living in areas of lead mineralization are exposed to higher levels of lead

Table 1.3 Major consumers of refined lead (1978) data from [2].

Country	Refined lead consumption (million tonnes)	Percentage of world consumption
USA	0.976	22
USSR	0.620*	14
Germany FR	0.272	6.1
Japan	0.266	6.0
UK	0.242	5.5
France	0.212	4.8
China PR	0.200*	4.5
Italy	0.191	4.3
Others (all <0.100)	1.451	33
Total	4.430	

*Estimated.

Table 1.4 Principal end uses of refined lead* in western world countries [2].

	Quantity (million tonnes)			
	1964	1969	1974	1977
Batteries	0.772	0.992	1.390	1.478
Pigments and chemicals	0.248	0.289	0.360	0.369
Tetraalkyllead	0.254	0.319	0.317	0.292
Cable	0.427	0.352	0.322	0.216
Pipe and sheet	0.298	0.266	0.173	0.160
Miscellaneous	0.509	0.499	0.546	0.480
Total	2.508	2.717	3.108	2.995

*Including direct use of lead contained in scrap.

than the general population. Recently, however, concern has centred upon pollution arising from the use of lead additives in gasoline (petrol), which cause emissions of lead aerosol when burned.

People are exposed to lead through breathing lead-polluted air and through the ingestion of lead in food and drink. The relative importance of each particular route of exposure is a matter of some uncertainty, and does of course vary considerably between individuals dependent upon their places of residence and work, and their dietary habits. Lead has always been present at low levels in food and drink due to the natural occurrence of the metal in rocks and soils. Deposition of lead aerosol may enhance the levels of lead in foodstuffs and water, but the magnitude of this is difficult to quantify.

Although low-level exposure is tolerated, lead acts as a poison when taken into the body in sufficient quantity. Recognizable symptoms of poisoning are known to occur in some individuals when levels of lead in the bloodstream exceed 70–80 μg(Pb) per 100 ml of whole blood. In comparison, typical levels of lead in the blood of the general population lie between 10 and 30 μg(Pb) per 100 ml. Over recent years, however, there has been mounting concern over possible, more subtle adverse effects to health at lower levels of exposure, and hence blood lead levels, than previously thought safe. Concern has been directed particularly at possible 'sub-clinical' injury to children. Such injury is believed to take the form of slight, but irreversible damage to brain development in the growing child. As such, it is very difficult to detect or quantify, and its existence is hotly disputed.

Basing their views upon an examination of the biochemical evidence for adverse effects of lead rather than upon studies of child brain development, the US Environmental Protection Agency (EPA) [4] and the World Health Organization (WHO) [5] have recommended upper limits to the population median blood lead level. In consequence, the US EPA has promulgated an ambient air quality standard designed to limit the exposure of the general population to airborne lead [4]. This standard will necessitate tight control over emission of lead into the atmosphere.

Human exposure to lead occurs through air, water and foodstuffs. Since the passage of lead into and between these media involves many complex environmental pathways, any appraisal of control techniques for lead in the environment must take account of these pathways. Hence this book aims to introduce the reader to the environmental science of lead, as well as to the techniques of engineering control. In view of the current concern and legislation over lead, this technology will be of increasing importance in the future.

References

[1] Grandjean, P. (1975), Lead in Danes. Historical and Toxicological Studies, in *Environmental Quality and Safety, Suppl. Vol II, Lead*, (ed. F. Coulston and F. Korte), Academic Press, New York, San Francisco, London, pp. 6–75.

[2] World Bureau of Metal Statistics (1979), *World Metal Statistics*, Vol. 32, World Bureau of Metal Statistics, London, pp. 72–8.

[3] Robinson, I. M. (1978), Lead as a Factor in the World Economy, in *The Biogeochemistry of Lead in the Environment*, (ed. J. O. Nriagu), Elsevier/North-Holland Biomedical Press, Amsterdam, pp. 99–118.

[4] US Environmental Protection Agency (1978), National Ambient Air Quality Standard for Lead, *Federal Register*, **43**, 46246–77.

[5] World Health Organization (1978), *Health Hazards from Drinking-Water*, Report on a Working Group, London, September 1977, WHO, Copenhagen.

2

Lead in the atmosphere

2.1 Introduction

There are two very important reasons for studying lead in air. Firstly, inhaled lead may pass via the respiratory system to the bloodstream and hence contribute to the lead exposure of the population. Secondly, airborne lead is progressively removed from the atmosphere by wet and dry deposition processes causing contamination of other environmental media.

2.2 Sources of lead in air

Emissions of lead into the atmosphere were estimated by the US Environmental Protection Agency (EPA) for the year 1975, and these data appear in Table 2.1. Clearly, by far the greatest source is emission from vehicles burning leaded petrol, although the relative importance of this source in the US may have diminished in recent years due to the progressive introduction of lead-free petrol, and in some other countries due to reductions in the lead content of petrol. The lead industry is important as a source of lead in air but on a more local scale than vehicle emissions.

2.2.1 Industrial emissions

Smelting and refining of lead give rise to a considerable generation of fumes. Emission factors for total particulates from processes involved in the production of primary and secondary lead are listed in Table 2.2. Modern control devices have an extremely high efficiency, and emissions into the atmosphere are greatly restricted. Hence in 1971, total particulate emissions into the atmosphere from production of primary and secondary lead in the US were estimated to be 30 000 and 4000 tonnes per year respectively [2]. The lead content of such emissions is estimated by Lee and von Lehmden [3] to be in the range 0.1–100 mg kg^{-1}, although in our own experience it is very much higher than this, and hence emissions of lead itself are considerably less (cf. Table 2.1). UK emissions of lead from smelting and refining were estimated at 200–250 tonnes y^{-1} in 1974 [4]. The lead content of particulate emissions from other industrial processes is summarized in Fig. 2.1.

Table 2.1 US national atmospheric lead emissions in 1975 [1].

Process	Emission (tonne)	Emission (US tons)
Gasoline combustion	127 800	140 900
Coal combustion	228	257
Oil combustion	100	110
Solid waste incineration	1 170	1 296
Waste oil disposal	5 000	5 480
Lead alkyl production	1 000	1 100
Storage battery production	82	90
Ore crushing and grinding	493	544
Primary lead smelting	400	440
Primary copper smelting	1 314	1 444
Primary zinc smelting	112	124
Secondary lead smelting	750	830
Brass and bronze production	47	52
Gray iron production	1 080	1 192
Ferroalloy production	30	33
Iron and steel production	605	667
Lead oxide production	100	110
Pigment production	12	13
Cable covering	113	125
Can soldering	63	70
Type metal	435	480
Metallic lead products	77	85
Cement production	312	344
Lead glass production	56	62
Total	141 380	155 880

Table 2.2 Emission factors for total particulates from lead smelting [2].

Production methods	(kg tonne^{-1})
Primary lead production	
1. Ore crushing	1 of ore
2. Sintering	260 of lead
3. Blast furnace	125 of lead
4. Dross reverberatory furnace	10 of lead
5. Materials handling	2.5 of lead
Secondary lead production	
1. Pot furnaces	0.5 of scrap
2. Blast furnaces	95 of scrap
3. Reverberatory furnaces	50 of scrap

Lead is inevitably released into the atmosphere during chemical processes in which it is used. For example, manufacture of lead alkyl antiknock compounds causes emission of both organic and inorganic compounds of lead, the losses of the former during manufacture and transfer contributing 140 tonnes of lead y^{-1} to the UK atmosphere in 1974 [4].

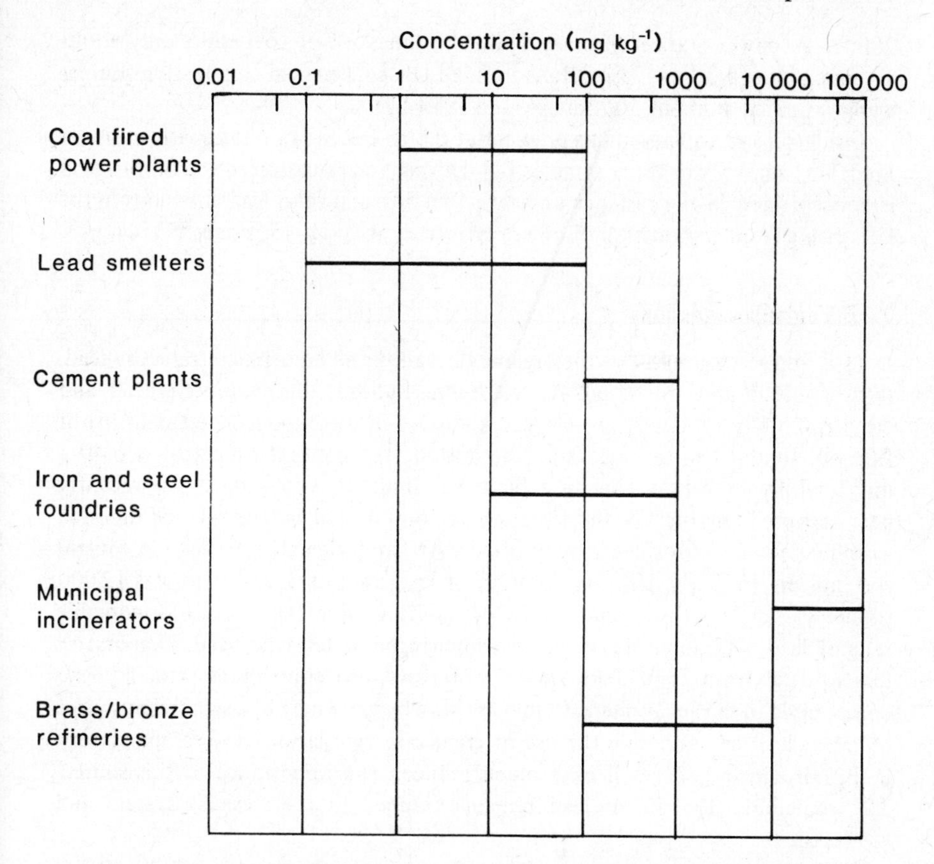

Fig. 2.1 Range of lead concentrations in particulate emissions. Samples collected isokinetically past control devices. (Based on [3]).

2.2.2 Fossil fuel combustion

The lead content of US coal is 35 ± 44 mg kg^{-1}, and in the UK, 17 mg kg^{-1} on average [4]. During combustion, most lead becomes concentrated in the ash, where mean concentrations may reach 200 mg kg^{-1} [4]. In domestic coal burning, chimney gas velocities are low, and the majority of ash remains in the grate. In commercial combustion, such as a power station, however, a far higher proportion of ash is entrained as fly ash and the amount reaching the atmosphere is highly dependent upon the control devices in use. Lee and von Lehmden measured concentrations of lead in the range 1–100 mg kg^{-1} of lead in power station emissions, with a fairly uniform concentration in each particle size range collected [3]. The arrestment efficiency of modern electrostatic precipitators is so high that the UK Department of the Environment estimates that a modern

2000 MW power station burning 5×10^6 tonnes y^{-1} of coal emits only about 1 tonne y^{-1} of lead [4]. Emissions from all UK commercial combustion sources were estimated at about 50 tonnes y^{-1} in 1974 [4].

Crude oil typically contains only about 0.001-0.2 mg kg^{-1} lead with an upper limit of 2 mg kg^{-1} in some samples [4]. During fractionation of the oil this lead is concentrated in the residues or heavy fuel oils, and total lead emissions in the UK arising from combustion of oil are estimated at about 15 tonnes y^{-1} [4].

2.2.3 Vehicular emissions

Lead is added to gasoline as the organic tetraalkyllead additives tetramethyllead, tetraethyllead and mixed alkyls triethylmethyllead, diethyldimethyllead and ethyltrimethyllead. There are no lead additives in diesel, kerosene (paraffin) or fuel oil. In the UK the maximum permitted lead content of petrol is 0.40 g dm^{-3}, whilst in West Germany a far lower limit of 0.15 g dm^{-3} has recently been adopted. In the US the situation is complicated by the sale of unleaded gasoline necessary for the cars now fitted with catalytic exhaust emission control systems. In 1973 the UK consumption of lead antiknock additives was 12 000 tonnes y^{-1} [4]. Of this, about 70-75% is emitted from the exhaust as inorganic salts of lead, and about 1% is emitted unchanged as tetraalkyllead. Evaporative loss of fuel from fuel tanks and carburettors also contributes tetraalkyllead compounds to the atmosphere. Crankcase blowby gases may be a significant source of tetraalkyllead, although the use of crankcase ventilation devices, mandatory in Britain since 1972, will have much reduced the importance of this source. The remaining 20-25% of lead remains trapped in the exhaust system and engine oil.

Concentrations of lead in vehicle exhaust gases are typically 2000-10 000 $\mu g(Pb) m^{-1}$. The quantity of inorganic lead emitted as a proportion of that consumed by the engine is highly dependent upon the driving mode [5], as illustrated in Fig. 2.2. The vast variation, from less than 5% during city-type driving to almost 2000% during rapid acceleration, is due to the deposition and subsequent resuspension of lead from the vehicle exhaust system. The emission of tetraalkyllead is also dependent upon driving mode, with from <0.1 to 5% of input lead emitted in this form.

The organic tetraalkyllead compounds are volatile and exist in air in the vapour phase, whilst the inorganic salts are emitted as particles. These particles have a very wide range of sizes. Research work has indicated that there are two main size groupings of particles in exhaust emissions: Those of $<1 \mu m$ aerodynamic diameter, and those of 5-50 μm. The former particles represent the primary lead emission, whilst the latter are presumably those resuspended from the exhaust system. A small fraction of larger 'gritty' lead-containing particles of 300-3000 μm diameter has also been noted by one worker [6]. More recent research has shown that the primary exhaust particulates are approximately 0.015 μm in diameter [7], and these particles will be subject to rapid growth

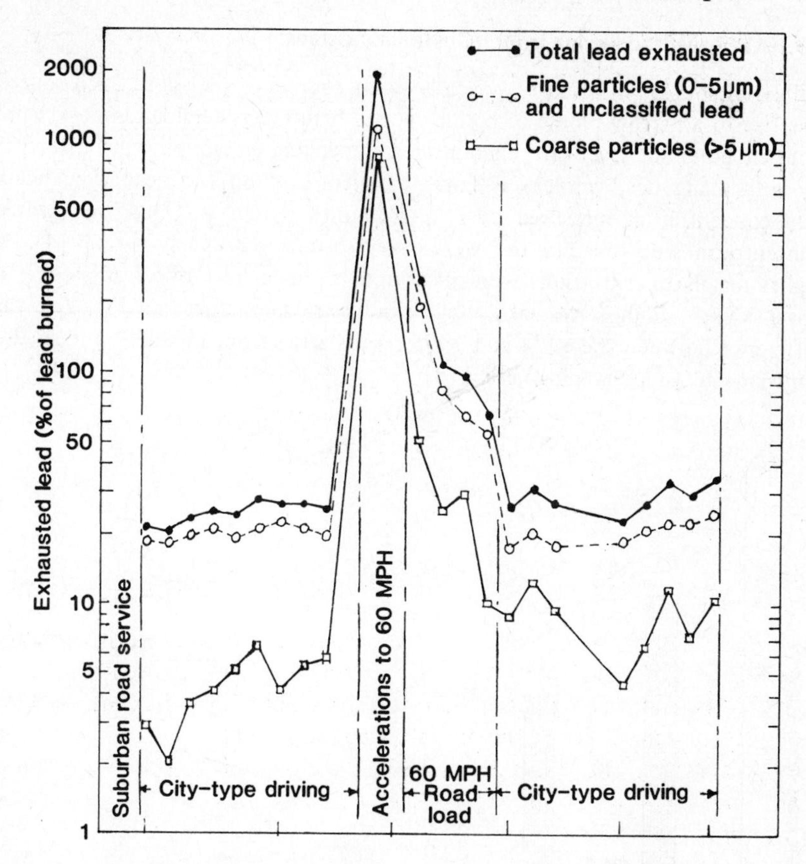

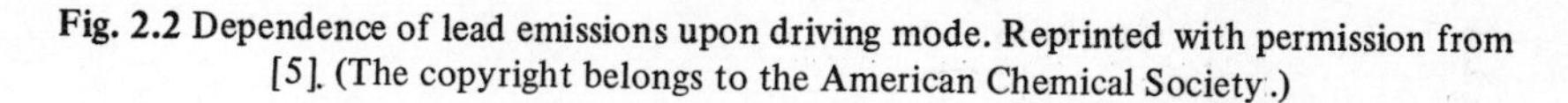
Fig. 2.2 Dependence of lead emissions upon driving mode. Reprinted with permission from [5]. (The copyright belongs to the American Chemical Society.)

in the ambient air by coagulation with other particles due to their high Brownian diffusivity.

2.3 Concentrations of lead in ambient air

There is no simple relationship between the source strength and ambient concentration of an atmospheric pollutant, since dispersion of the pollutant between source and receptor is a process which is dependent upon meteorological conditions and is hence very variable. For this reason, concentrations measured over a short time period at only one site may not be representative of ambient concentrations in that region, and longer-term average concentrations measured at carefully selected sites are to be greatly preferred.

2.3.1 Airborne lead arising from industrial emissions

Industrial sources of lead are typically point sources of emission. Such sources give rise to a narrow plume of pollution, and hence considerable temporal variations in pollutant levels are encountered dependent upon the wind direction. Close to a large lead refinery at Tower Hamlets, London, the mean atmospheric lead concentration measured over a six month period was 3 μg m^{-3}, whilst concentrations up to 92 μg m^{-3} were observed during 3 hourly periods [4]. Far higher mean concentrations were measured in the vicinity of a lead smelter in Silver Valley, Idaho. Annual average results are summarized in Fig. 2.3. The difference between the 1974 and 1975 results arises from a reduction in smelter emissions in the intervening period [8].

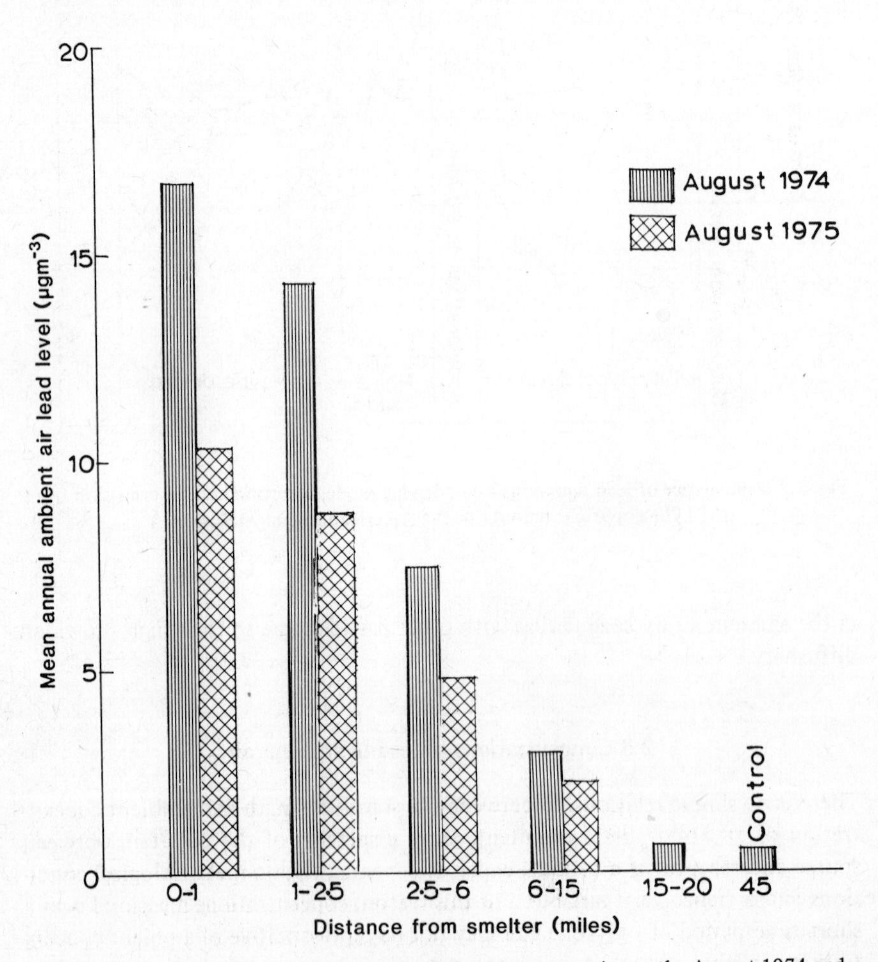

Fig. 2.3 Annual ambient air lead concentrations, by area, prior to the August 1974 and August 1975 surveys [8].

2.3.2 Airborne lead arising from vehicular emissions

As noted earlier, lead emissions from gasoline-fuelled motor vehicles take the form of both organic and inorganic lead. Since both the environmental and toxicological properties of these two chemical forms of lead are rather different, they will be considered separately.

2.3.2.1 Inorganic lead in air

By far the major proportion of vehicle-emitted lead in air is inorganic, and hence exists in the air in particulate form. The fate of particulate pollutants is highly dependent upon the particle size.

Particles of >10 μm diameter are subject to significant sedimentation rates as a result of gravitational forces (the velocity may be estimated crudely by use of Stoke's Law). Consequently their atmospheric lifetime is severely limited by the gravitational settling process, and by impaction upon surfaces. Particles of <10 μm diameter are removed only relatively slowly from the atmosphere, those greater than *ca.* 0.3 μm by impaction processes and the smaller particles by diffusive deposition. The deposition flux may be estimated from the deposition velocity and some data are presented in Section 4.2.2.

Estimates of the atmospheric lifetime of lead aerosol range typically between 7 and 30 days. This time is sufficient for transport over thousands of kilometres, and consequently even the most remote sites may experience pollution by inorganic lead. It is hence valuable to consider inorganic lead pollution under two headings: roadside and urban (close to source) and rural (remote from source) pollution.

Roadside and urban concentrations

In an American study, mean air lead concentrations were related to traffic volume and distance from the highway [9] with the results shown in Fig. 2.4. The magnitude of such pollution will, of course, depend upon the source strength (traffic volume, concentration of lead in petrol, driving modes) and the prevailing meteorological conditions. It may be seen that lead concentrations fall rapidly with distance from the highway as a result of dispersion processes. Chamberlain and co-workers [7] give a predictive equation from which lead concentrations in the proximity of a roadway may be calculated. The concentration, $\chi(x)$, is given by

$$\chi(x) = \left(\frac{2}{\pi}\right)^{\frac{1}{2}} \frac{q}{uw} \int_{x}^{x+w} \frac{dx}{\sigma_z(x)} \qquad (1)$$

in which q is the source strength of lead in μg m^{-1} s^{-1}, u the mean wind velocity, w the width of highway in metres, x the distance of point of measurement from edge of highway in metres and $\sigma_z(x)$ the standard deviation in metres of the vertical spread of the plume which is equal to ax^s. a and s are parameters dependent upon atmospheric stability (i.e. the degree of turbulence of the atmosphere).

The derivation of this equation assumes that cars travel equally spaced in all

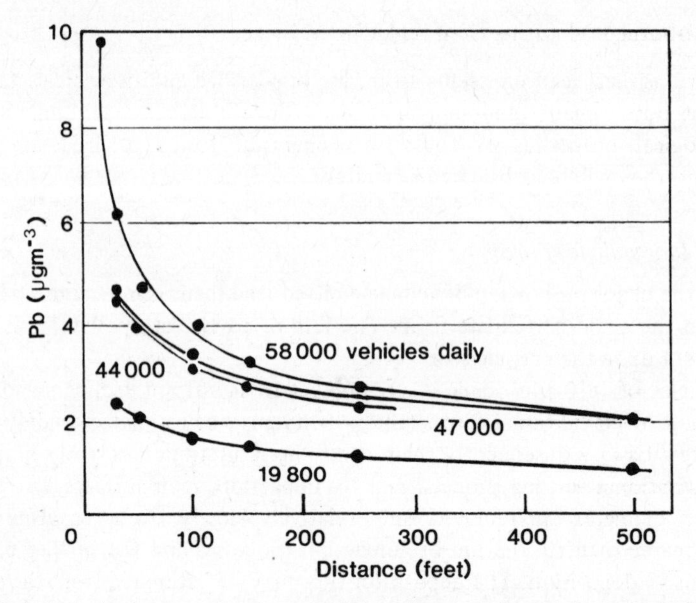

Fig. 2.4 Air lead values as a function of traffic volume and distance from the highway. Reprinted with permission from [9]. (Copyright by the American Chemical Society.)

lanes, and that the wind direction is perpendicular to the highway. The angle of the wind to the road has little effect upon lead concentrations until it exceeds 60° to the perpendicular. As the wind direction approaches parallel with the road, χ is increased and at an angle of 75° to the perpendicular, the increase is about 50% relative to a wind perpendicular to the road [7].

Using the above equation, Chamberlain *et al.* [7] have calculated ambient concentrations at a wind speed of 2 m s^{-1}, defining atmospheric stability in terms of the Pasquill categories B (unstable), D (neutral) and E (stable). The results appear in Fig. 2.5. These predictions have been shown to fit the measurements of lead concentration made alongside British motorways rather well [7]. Chamberlain estimates that as a generalization, at open sites with freely moving traffic the kerbside lead concentration is about 2 $\mu g\ m^{-3}$ per 1000 vehicles h^{-1} traffic. In narrow streets, with slow moving traffic the value is higher, up to 10.6 $\mu g\ m^{-3}$ per 1000 vehicles h^{-1} in a motorway cutting [7]. Selected roadside measurements made on busy highways or city streets appear in Table 2.3.

In an urban area, there are two separate effects. In addition to the lead arising from a nearby road, an urban background of perhaps 0.5–1 $\mu g\ m^{-3}$ of lead exists arising from the network of roads in the area. The effect of one particular highway becomes difficult to discern beyond about 100 m from the carriageway, and the lead concentration is best described in terms of emissions within the whole urban area.

Chamberlain and co-workers [7] have calculated concentrations of lead in air from area-wide sources. The concentration, C, near the ground is given by

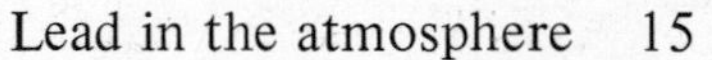

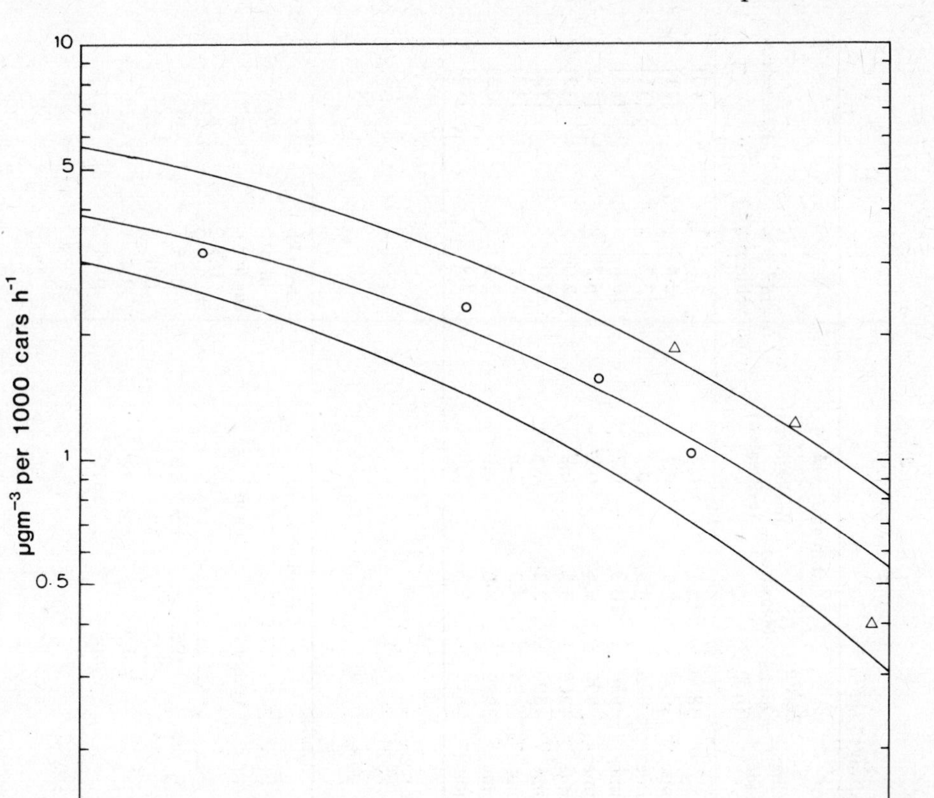

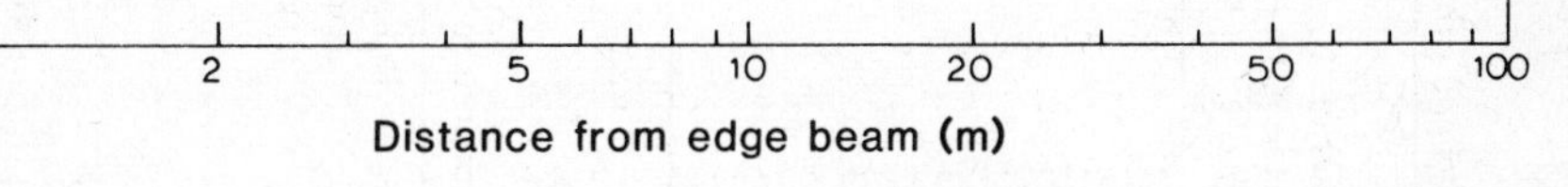

Fig. 2.5 Lead in air versus distance from motorway. Lines are labelled with Pasquill stability categories, [7]. Experimental measurements: ○, Little and Wiffen [11]; △, Bevan *et al.* [10] – see Table 2.3.

$$C = \int_{x_1}^{x_2} \left(\frac{2}{\pi}\right)^{\frac{1}{2}} \frac{Q}{u} \frac{dx}{\sigma_z(x)} \tag{2}$$

where the concentration is calculated for the centre of an urban area, radius x_2 with source strength Q in μg m^{-2} s^{-1}, and u and $\sigma_z(x)$ have the same meaning as in Equation 1. Hence,

$$\frac{C}{Q} = \left(\frac{2}{\pi}\right)^{\frac{1}{2}} \frac{(x_2^{\,1-s} - x_1^{\,1-s})}{u\,a\,(1-s)}, \tag{3}$$

Appropriate values of u, a and s are selected dependent upon meteorological

Table 2.3 Measurements of lead-in-air near motorways and other roadways [7].

Location	Open (O) or constricted (C) site	Sampling periods	Petrol vehicles (h^{-1})	Lead in air ($\mu g\ m^{-3}$) Measured	Lead in air ($\mu g\ m^{-3}$) Background (estimated)	χ ($\mu g\ m^{-3}$ per 1000 vehicles h^{-1})	References
M4, Central reservation	O	24 h	3800	15.1	0.4	3.9	[10]
M4, 2 m from edge beam	O	Daytime	3600	8.9	0.4	2.4	[11]
M40, 2 m from edge beam level	O	Daytime	970	3.3	0.2	3.2	[11]
1/20 grade in cutting	C	Daytime	950	10.3	0.2	10.6	[11]
1/20 grade on embankment	O	Daytime	990	6.4	0.2	6.2	[11]
Fleet St., London, E1, central	C	Daytime	1200	3.2	0.8	2.0	[12]
Exhibition Rd., London, SW 7, kerbside	O	Daytime	1300	3.2	0.5	2.0	[13]
Seymour Place, London, W1	C	Daytime	424	4.0	1.0	7.1	[13]
Talgarth Rd., London, W14	O	Daytime	3880	8.8	0.5	2.2	[13]
Upper Berkeley St., London, W1	C	Daytime	420	4.3	1.0	7.9	[13]

Table 2.4 Areal emissions of lead and concentrations in air.

Region	Area (km^2)	Equivalent diameter (km)	Emission of Pb (tonnes y^{-1})	Emission of Pb ($\mu g\ m^{-2}\ s^{-1}$) (Q)	Mean Pb in air ($\mu g\ m^{-3}$) (C)	C/Q ($s\ m^{-1}$)
Birmingham CB	2.1×10^2	16	125	1.9×10^{-2}	0.75	39
Inner London	2.3×10^2	17	270	3.8×10^{-2}	0.8	22
Los Angeles County	4.4×10^3	75	6400	4.6×10^{-2}	2.4	52
UK	1.7×10^5	465	9000	1.7×10^{-2}	0.14	82

Sources of data referenced in Chamberlain *et al.* [7].

conditions, and the lower limit of the integration, x_1, is taken as 50 m since the equation is designed to predict concentrations at sites not immediately adjacent to roads [7].

In Table 2.4 relevant data for selected areas are presented, and in Fig. 2.6 these are compared with the predictions of Equation 2 which have been calculated using time-weighted values of Q, and of the atmospheric stability category to take into account the diurnal variation in emissions and in atmospheric stability [7]. This is important since the atmosphere is frequently most stable at night when emissions are low, and as a consequence night-time concentrations are often as high, or higher than day-time values. It may be seen that the agreement between theory and practice is reasonably good.

Atmospheric concentrations measured at a range of sites in the US during 1974 are shown in Table 2.5. These show typical values measured at that time, although concentrations have probably decreased somewhat in the intervening period as a result of increasing usage of lead-free gasoline.

Rural concentrations

These are illustrated in Fig. 2.7 which shows a pollution rose constructed from data collected at rural sites in the region of Lancaster, England [15]. Very low concentrations (as low as 0.004 $\mu g\ m^{-3}$, 24h average) of inorganic lead are associated with air masses arriving via maritime trajectories, whilst the higher concentrations of up to 0.66 $\mu g\ m^{-3}$ are carried by air masses travelling over urbanized areas of Britain. Often the very highest concentrations are found in air masses having travelled over continental Europe before reaching the UK. Hence, transported aerosol is clearly the main source of lead at remote rural sites. Sampling downwind of an urban area, elevated concentrations of lead may be expected within the 'urban plume' of pollution carried by the prevailing wind. Natural particulate lead levels are estimated from geochemical evidence to be around 0.0005 $\mu g\ m^{-3}$ [4].

2.3.2.2. Organic lead in air

This may take two forms. The major part consists of vapour phase tetraalkyllead compounds which may be separated from inorganic lead by filtration of an air sample. A small portion of the organic lead exists in association with atmospheric particles, and probably takes the form of tetraalkyllead or trialkyllead salts adsorbed upon particulate air pollutants.

The scientific literature has shown a wide variety of concentrations for tetraalkyllead vapour in air. These are best considered as a percentage of total lead in air, since this allows comparison of data from different sites without the need to correct for the effect of topography and meteorological conditions at the time of sampling. Even these percentages vary widely, but Harrison and Perry [16] have argued that many of the disparities have arisen from the use of non-specific analytical techniques. Values appear far more consistent if

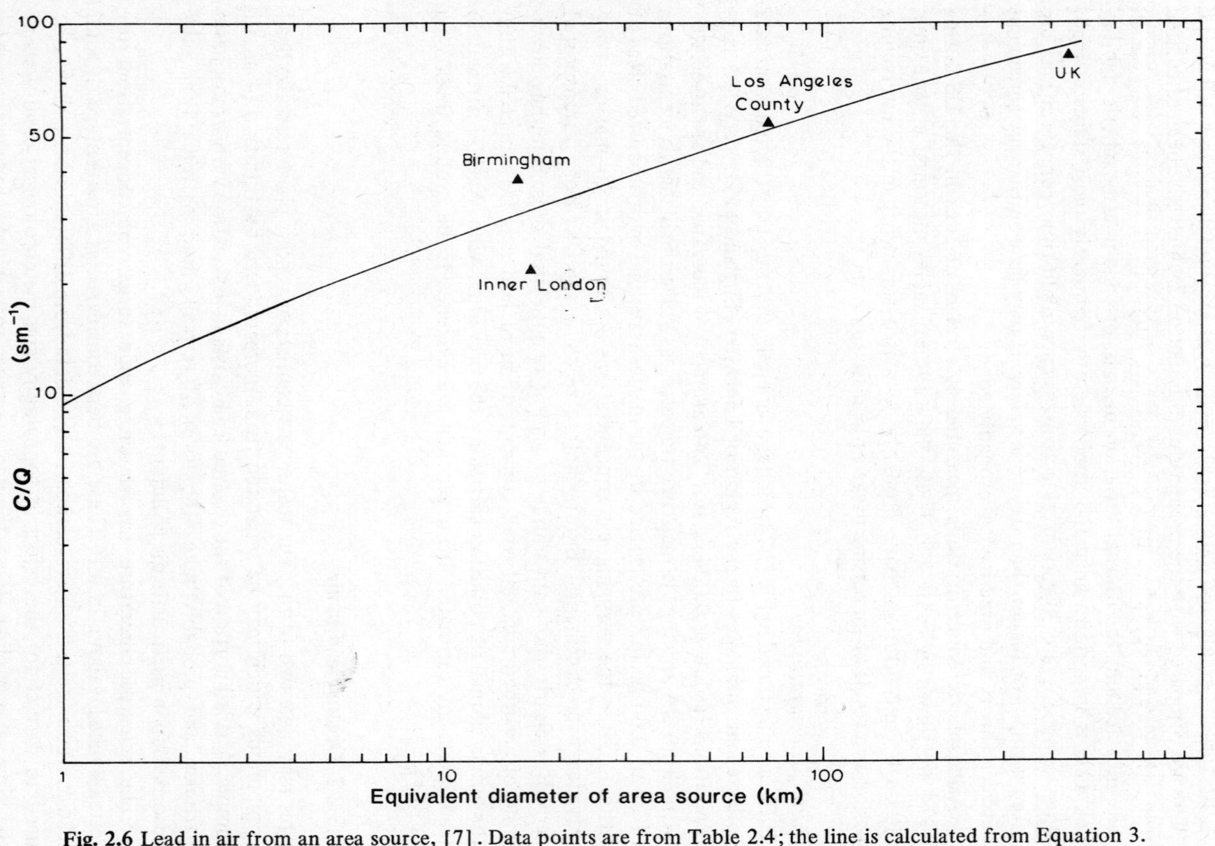

Fig. 2.6 Lead in air from an area source, [7]. Data points are from Table 2.4; the line is calculated from Equation 3.

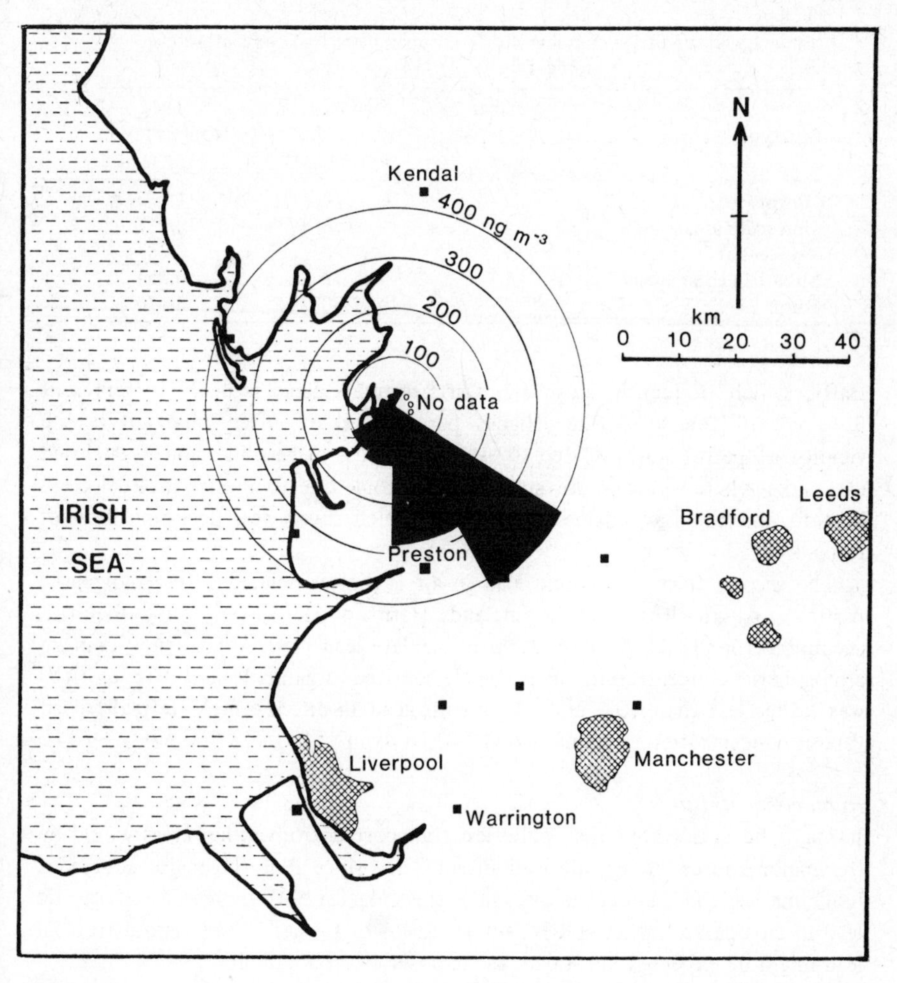

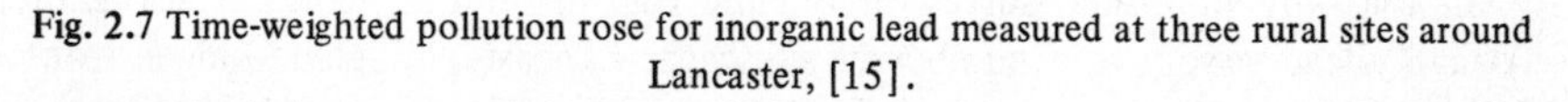

Fig. 2.7 Time-weighted pollution rose for inorganic lead measured at three rural sites around Lancaster, [15].

those results obtained with specific, reliable analytical techniques only are considered.

Roadside and urban concentrations

At most sites the typical contribution of vapour phase tetraalkyllead to total lead lies within the range 1–6%. Figures taken from a recent review article are given in Table 2.6. It has been pointed out that at roadside sites the chief determinant of tetraalkyllead as a percentage of total lead is the predominant vehicle driving mode [17]. Alongside British motorways where vehicles travel at high, constant speeds tetraalkyllead may account for as little as 1% of less total lead. At urban sites affected by stationary vehicles, cold choked vehicles or

Table 2.5 Ground-level sampling sites – mean ambient lead concentrations (May–July 1974) [14].

Site type	Number of sites	Number of observations	Lead concentration ($\mu g\ m^{-3}$)
Downtown	2	76	1.352
University area	4	158	0.747
Residential	9	342	0.287
Suburban commercial	1	39	0.254
Rural	2	80	0.170

traffic which frequently stops and starts, tetraalkyllead is likely to represent 5 to 6% of the total lead. Higher percentages are often observed in the vicinity of petrol stations, due to evaporative loss of gasoline, and in enclosed car parks where vehicles are started from cold, due to inefficient fuel combustion and the low engine temperature which causes less decomposition of tetraalkyllead.

The second form of organic lead in air is that associated with atmospheric particles. At an urban site in England, Harrison and Laxen found that this accounted for 0.2–1.2% of the total particulate lead [18]. At that site the mean atmospheric concentration of particle-associated organic lead (0.015 $\mu g\ m^{-3}$) was rather less than 10% of total organic lead in air, the mean tetraalkyllead vapour concentration being measured as 0.16 $\mu g\ m^{-3}$ [19].

Rural concentrations

It might be anticipated that pollution transported from urban areas would be the major source of organic lead in rural air, since it is the major source of inorganic lead. This being the case, since the sinks for both inorganic and organic lead in air operate rather slowly, similar ratios of tetraalkyllead vapour to total lead might be expected in rural air as are observed in urban air.

In the only published survey of organic lead in rural air, concentrations of tetraalkyllead vapour at a number of sites near Lancaster, England ranged from 0.5 to 230 $ng\ m^{-3}$ representing from 1.5 to 33% of total lead [15]. A time-weighted pollution rose appears in Fig. 2.8. Neither the higher concentrations nor the higher percentages could be explained by advection from urban areas. Calculation of time-weighted percentages of tetraalkyllead/total lead related to wind direction, shown in Fig. 2.9 revealed low percentages to be associated with overland winds and the higher percentages to be associated with maritime winds. Similarly when geostrophic air mass back-trajectories were calculated the lowest daily percentages of about 2% were found to be associated with air having passed over urban source areas. On days when tetraalkyllead percentages exceeded 20%, the air mass had arrived over the sea, typically also travelling over intertidal areas of Morecambe Bay close to the sampling sites. It was concluded that in the Lancaster area, natural environmental methylation of lead probably occurring

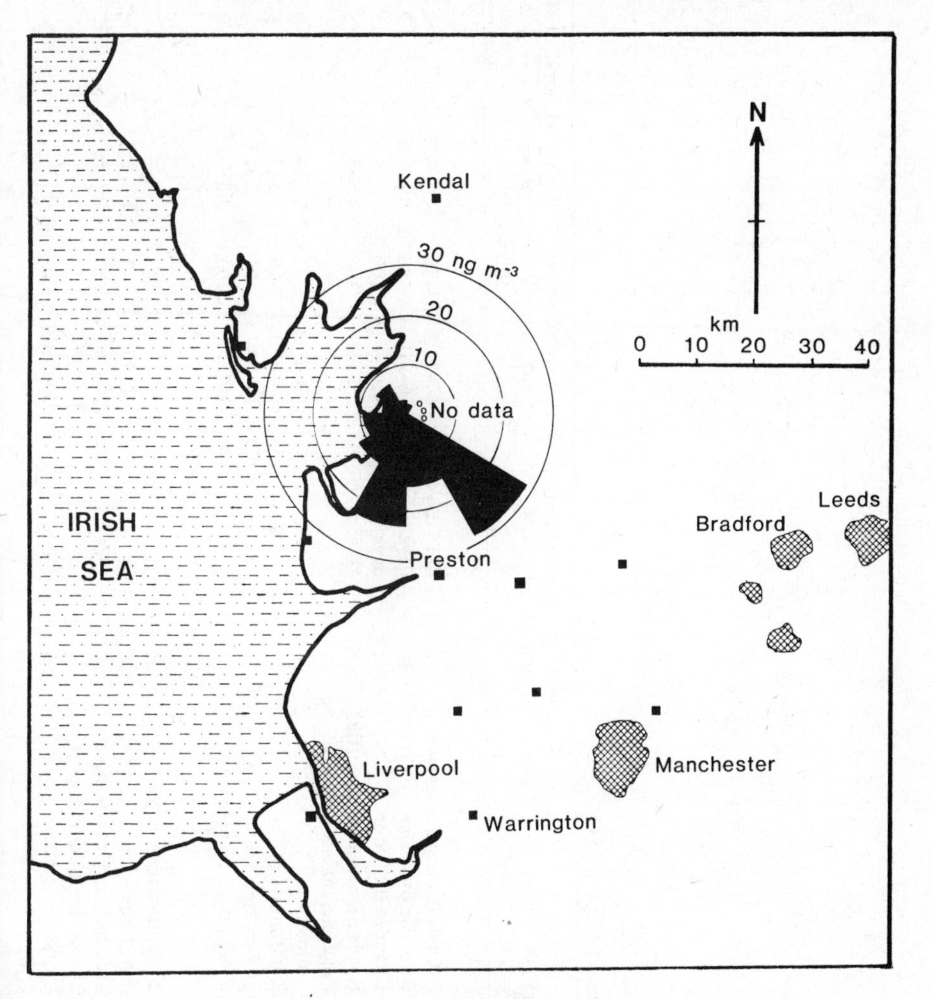

Fig. 2.8 Time-weighted pollution rose for tetraalkyllead vapour measured at three rural sites around Lancaster, [15].

within the bay area must contribute tetramethyllead to the atmosphere, hence causing the elevated percentages associated with certain wind directions and air mass trajectories. The generality of this observation is not known. It does however seem probable that at inland rural sites tetraalkyllead percentages will be similar to those at urban sites, whilst other factors may affect coastal sites causing elevated tetraalkyllead vapour concentrations in the atmosphere.

2.3.2.3 Vertical profiles of lead above a street

Often, city streets form a 'canyon' from which dispersion of pollutants occurs only rather inefficiently. In a recent study [20] both inorganic and vapour-phase

Table 2.6 Measurements of tetraalkyllead (TAL) in air [17].

Location	Number of sites	Type of site	Number of samples	Averaging time	TAL concentration (ng(Pb)m^{-3}) Range	Mean	Ratio TAL/total Pb(%) Range	Mean
Averaging time – days								
Los Angeles	1	300 yds from major highway	6	4–15 days	47–110	78	1.5–4.0	2.4
Delaware	1	Roadside by bridge	10	days	< 6–210	83	0.3–1.8	0.8
Los Angeles	1	Urban air 3rd floor	1	days		100		5
Lancaster (UK)	1	Roadside, accelerating traffic	3	10 days	75–260	160	4.5–5.7	5.0
Shap	1	Roadside, remote section of motorway	4	1 day	0.8–2	1.4	0.22–0.46	0.33
Lancaster (UK)	4	Rural sites	33	1–2 days	0.5–230	19	1.5–33	9.5
London	1	City centre street at (1)	7	1 day	(1) 24–190	94	3.2–9.8	6.2
		4.9 m, (2) 14 m heights	–	–	(2) 16–130	65	3.5–13	6.7
Averaging time <1 day								
Baltimore	1	20 m downwind major highway	6	2h	26–75	53	1.5–5.2	3.2
Stockholm	3	Wide busy city centre streets	6*	1–24 h†	220–950	360‡	5.7–19	12
Stockholm	2	City centre (a) crossing (b) queue of traffic	5*	3–8 h†	120–1300	510	–	–

Frankfurt	5	From rural-city centre	104	2 h	1–170 §		0.3–24	–
London	5	Roadside, on kerb	6	30 min	30–110	68	0.9–4.1	2.1
London	1	Roadside, 5 m from kerb	7	30–40 min	60–200	130	3.0–7.9	4.5
	1	Works exit traffic lights,						
		traffic (1) arriving (2)	3	20 min	(1) <100–300	<200	–	–
		leaving	9		(2) <100–2500	<770	–	–
	1	Tunnel	9	2 h	57–130	92	0.37–0.83	0.55
	1	Tunnel	1	30 min		20		0.1
	1	Petrol station	2	30 min	240–590	420	3.9–9.7	6.8
	1	Petrol station	3	30–40 min	470–820	650	19–29	24
	1	Multi-storey car park	2	30 min	1500–5400	4200	10–13	11
	1	Car park	3	2–5 h	540–3400	1900	–	–
	1	Basement car park	4	2 h	500–1000	850 §	–	30

*Number of sampling periods.

†Each sample period made up of sequential 10–15 min samples. Individual samples show much wider range of concentrations.

‡At the time the max concentration of lead in petrol was 0.7 g dm^{-3}.

§Maximum lead in petrol 0.15 g dm^{-3}.

Sources of data are cited by Harrison, Laxen and Birch [12].

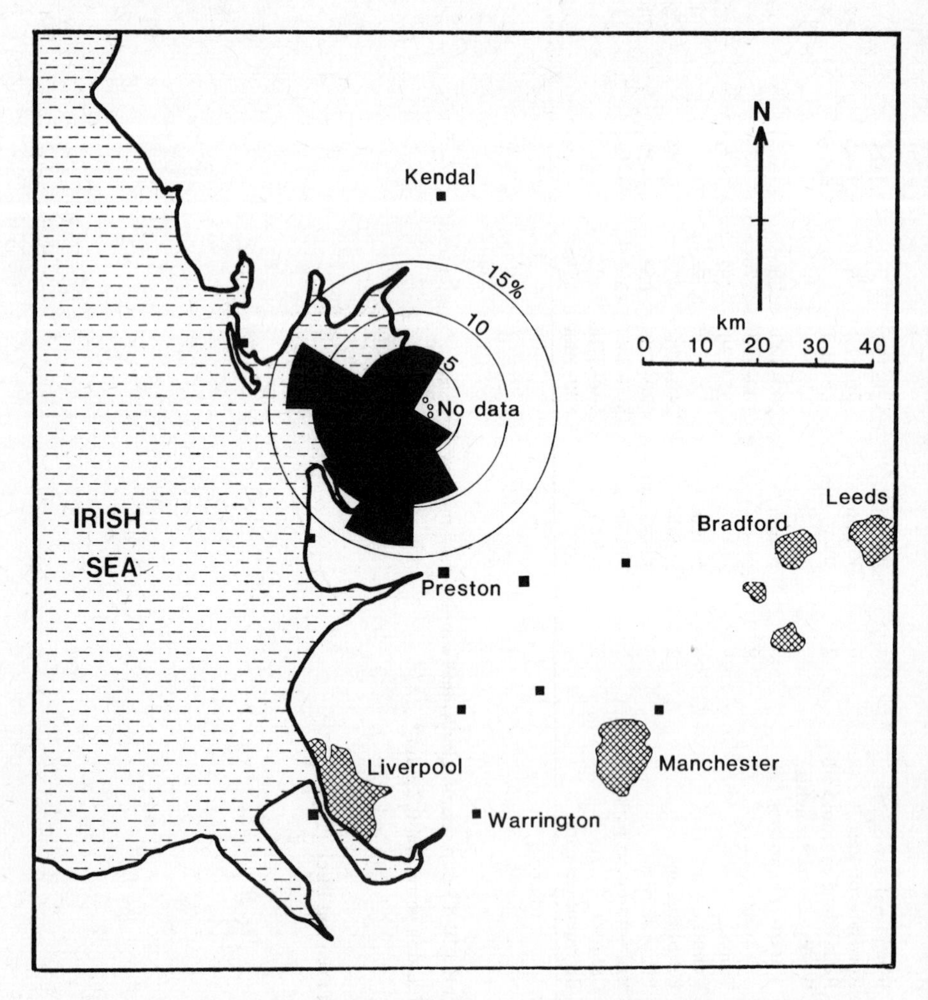

Fig. 2.9 Time-weighted pollution rose for the percentage ratio tetraalkyllead/total lead measured at three rural sites around Lancaster, [15].

organic lead were measured at two heights above a busy London street. The air was sampled at heights of 4.9 m and 14 m from first and fourth floor windows in vertical alignment. The traffic in the street below was frequently slow moving, with a pedestrian crossing to bring traffic to an occasional halt.

The results (Table 2.7) show the effect of strong winds in causing a rapid dispersion and dilution of fresh emissions and hence low concentrations on the first two days of the survey. There is a mean decrease of 35±8 (standard deviation, s.d.) % in particulate lead and 31±9 (s.d.)% in tetraalkyllead in going from 4.9 to 14 m above the street. The small increase in tetraalkyllead/total lead ratio with height may arise from slight loss of particulate lead during dilution and vertical transport, but was not found to be statistically significant.

Table 2.7 24 h concentrations* of particulate lead and tetraalkyllead measured in air above a central London street, [20].

Date	Floor	Particulate Pb ($\mu g\ m^{-3}$)	Tetraalkyllead ($\mu g\ m^{-3}$)	Total ($\mu g\ m^{-3}$)	% TAL/total lead	Wind direction and speed† ($m\ s^{-1}$)
11–12 Dec 1978	Lower‡	0.50	0.027	0.53	5.1	SE 6.6, SE 6.1, S 11.8, S 5.1, SW 7.2
	Upper	0.31	0.019	0.32	5.9	
12–13 Dec 1978	Lower	0.40	0.024	0.42	5.6	SW 7.2, S'W 6.1 SW 8.2, S'W 7.7, SW'W 7.2
	Upper	0.27	0.016	0.28	5.7	
18–19 Dec 1978	Lower	3.9	0.131	4.05	3.2	Calm, ENE 0.5, Calm, Calm, NNE 2.6
	Upper	2.7	0.095	2.75	3.5	
5–6 Feb 1979	Lower	1.9	0.134	2.06	6.5	NE 3.6, NE 5.6, E 3.1, NE 1.0, E'N 0.5
	Upper	1.5	0.101	1.62	6.2	
6–7 Feb 1979	Lower	1.8	0.192	1.95	9.8	E'N 0.5, SE 6.1, E'S 3.6, E 6.1, E'N 3.1
	Upper	0.88	0.132	1.02	12.9	
26–27 Feb 1979	Lower	1.0	0.046	1.07	4.3	SW 3.1, WSW 4.1, SW 3.6, SW 2.6, SW 3.6
	Upper	0.67	0.037	0.70	5.3	
27–28 Feb 1979	Lower	1.0	0.104	1.13	9.2	SW 3.6, SW 5.1, SSW 6.6, SSW 7.2, SE 4.6
	Upper	0.64	0.053	0.69	7.7	
Mean ± s.d.	Lower	1.5 ± 1.2	0.094±0.064	1.6±1.2	6.2±2.4	
	Upper	0.99±0.8	0.065±0.045	1.1±0.9	6.7±3.0	
Median	Lower	1.16	0.072	1.24	5.8	
	Upper	0.74	0.050	0.79	6.3	

*Each reported result is the mean of duplicate samples.

†Wind data from Kew at 6 h intervals.

‡Lower floor sample at 4.9 m. Upper floor sample at 14 m.

Table 2.8 Relationships between lead and carbon monoxide concentrations.

Reference	Sampling methods	Sampling position	Mean lead concentrations ($\mu g\ m^{-3}$)	Mean CO concentrations (ppm)	Number of pairs of measurements	Correlation coefficient (r)	Significance of correlation (%)	Equation of regression line
[21]	Lead: Hi-Vol sampler, glass fibre filters, 1.5 $m^3\ min^{-1}$ CO: Continuous	3 m from kerb, 0.5 m above ground; approximately 1000 vehicles h^{-1} in rush hour	0.5–1.6 (monthly)	Not given	78	0.72	99	Pb($\mu g\ m^{-3}$) = 0.63 CO (ppm) + 1.3
[22]	Lead: Millipore molecular sieve filter paper: 0.01 $m^3\ min^{-1}$ CO: Indicator tubes	10–17 m from traffic, approximately 1 m above ground (6 sites); 60–4140 vehicles h^{-1} average	< 0.1–4.74 (8 h)	<1–16 (8h)	30	Not given	Not given	Pb($\mu g\ m^{-3}$) = 0.268 CO (ppm) + 0.516
[23]	Lead: Sequential tape sampler, Whatman No. 4 paper: 0.007 $m^3\ min^{-1}$ CO: Continuous NDIR	2.6 m from kerb, 6 m, 14 m and 30 m above ground (3 sites); Heavy traffic	7.5	13.0	Correlation – no figures given.			
[24]	Lead: All-glass 'absolute' filter; 0.5–5.0 $m^3\ min^{-1}$ CO: Continuous	Various sites (12 in all)	0.4–18.4 (10–731 h)	1.5–19.3 (10–731 h)	111 (All sites, all data)	0.80	99	Pb($\mu g\ m^{-3}$) = 0.81 CO (ppm) + 0.95
[25]	Lead: Whatman No. 41 cellulose filters; 0.028 $m^3\ min^{-1}$ CO: Hourly measurement	Various sites (9 in all)	1.32–1.58 (96 h)	1.4–2.8 (96 h)	7 (Only 6 used in calculating regression line)	0.88	98	Pb($\mu g\ m^{-3}$) ≃ 0.5 CO (ppm)

2.3.2.4 Correlation of lead levels with other pollutants

The question of the correlation of lead concentrations with those of other vehicle-generated pollutants has been investigated by a number of research workers. The results of correlation with carbon monoxide are summarized in Table 2.8. It may be seen that significant correlations were found when long averaging periods were used, with different regression equations reported from each study. In reviewing this work, Perry and co-workers [26] point out that each sampling site will have its own characteristics in terms of topography, micro-meteorology and typical driving mode, hence, although a correlation of CO and lead may be observed, the regression equation is applicable only to that one site.

In a study in which correlations between short-term (10–25 min) measurements of lead, carbon monoxide and hydrocarbons were investigated, significant correlations between CO and both total hydrocarbons and gasoline vehicle traffic were observed within a multistorey car park. No significant correlations were observed for lead, however, at this and other urban highway sites [26]. Hence it is only over longer time periods where short-term variabilities are averaged that significant correlations are found.

2.4 Particle size distributions

Since the sizes of atmospheric particles are normally estimated from aerodynamic properties, these are frequently quoted as an aerodynamic equivalent diameter which is the diameter of an aerodynamically equivalent spherical particle of unit specific gravity. This convention is convenient since the behaviour of particles of $>0.1\ \mu m$ diameter in the atmosphere and in the human respiratory system is primarily dependent upon the aerodynamic size. The largest vehicle-emitted particles of 300–3000 μm diameter are subject to rapid gravitational settling, whilst the 5–50 μm fraction also settles close to source due to turbulent deposition upon surfaces [27]. The smaller particles of $<5\ \mu m$ have a far longer atmospheric lifetime, and it is mainly these particles which reach atmospheric sampling equipment, and which present the greatest hazard to health. They are eventually removed from the atmosphere by dry deposition upon surfaces and by precipitation scavenging processes known as rainout and washout.

As indicated earlier, primary exhaust lead particulates are of approximate diameter 0.015 μm and are subject to growth by coagulation processes [7]. In a polluted urban atmosphere, growth is rapid and continues until particle sizes of *ca.* 0.5 μm are reached by which time the Brownian diffusion mechanism of particle collisions becomes unimportant. Typical size distributions measured with Andersen impactors, which determine aerodynamic particle sizes, are shown in Fig. 2.10.

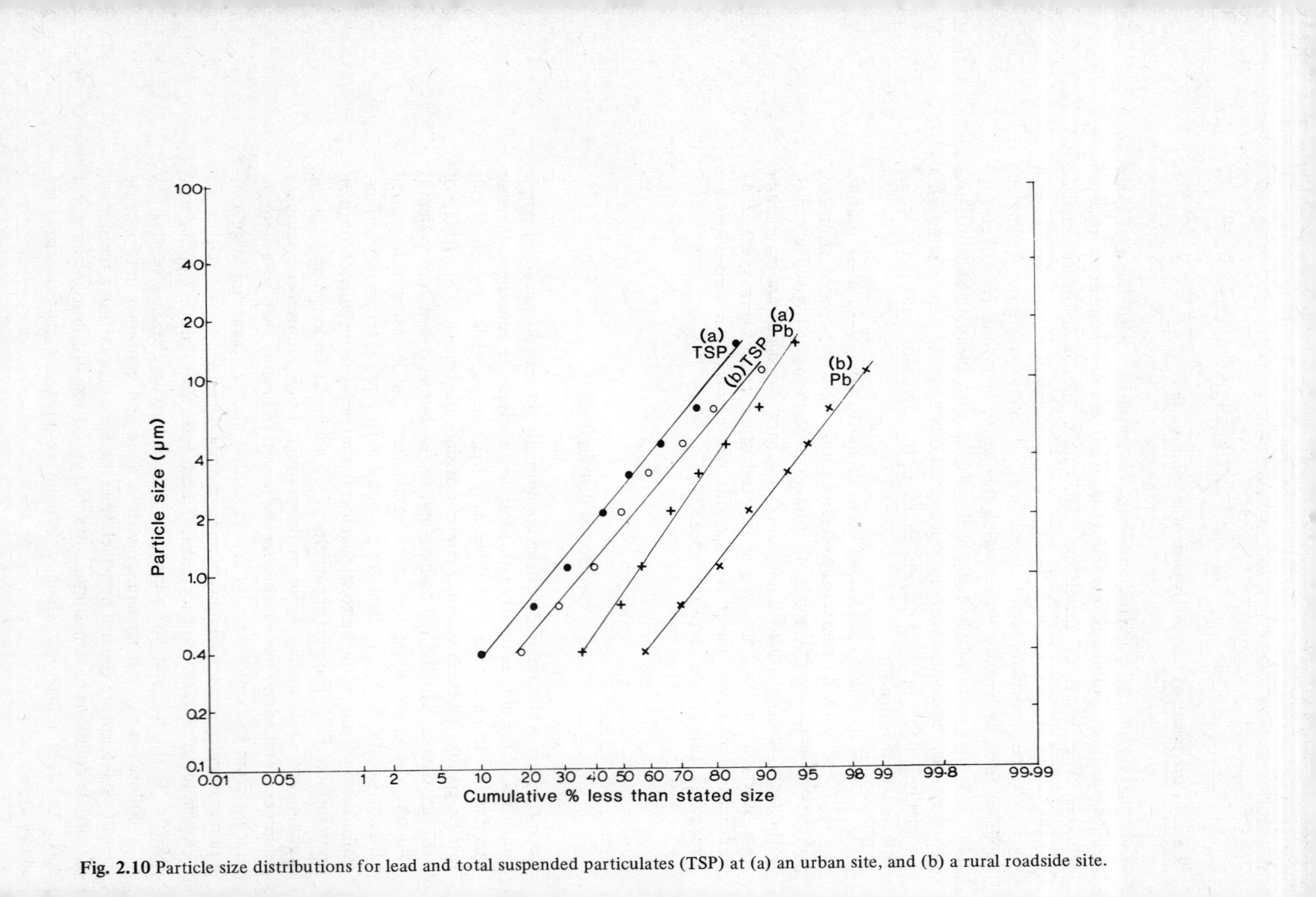

Fig. 2.10 Particle size distributions for lead and total suspended particulates (TSP) at (a) an urban site, and (b) a rural roadside site.

2.5 Chemical properties of atmospheric lead

The environmental mobility of lead is highly dependent upon its chemical form and hence the atmospheric chemistry of lead is a matter of some importance. In addition, the toxicity of lead is related to the chemical compounds to which the individual is exposed (Chapter 7).

2.5.1 Smelter emissions

Lead compounds are emitted from both the sintering and smelting processes during primary lead production (Chapter 5). Measurements both within the stack and the atmosphere in the vicinity of primary lead smelters indicate emission of lead primarily as $PbSO_4$ and $PbO{\cdot}PbSO_4$.

2.5.2 Vehicular inorganic lead

Several studies of vehicle exhaust particulates have revealed a complex mixture of compounds. The compounds identified are as follows [5, 28]

Compound	
$PbBrCl$	Major components
α-$2PbBrCl{\cdot}NH_4Cl$	
β-$2PbBrCl{\cdot}NH_4Cl$	Minor components
$PbBrCl{\cdot}2NH_4Cl$	
$3Pb_3(PO_4)_2{\cdot}PbBrCl$	
$PbSO_4$	
$PbO{\cdot}PbBrCl{\cdot}H_2O$.	

The major compound, lead bromochloride ($PbBrCl$) arises from reaction of lead oxide, formed by combustion of the tetraalkyllead additives, with HCl and HBr which arise from combusion of 1, 2-dichloroethane and 1, 2-dibromoethane [also known as ethylene dichloride (EDC) and ethylene dibromide (EDB)] which are added as scavengers in the fuel. The additives are most commonly present in a mole ratio of tetraalkyllead : EDB : EDC of 2 : 1 : 2 and hence an excess of HCl is formed which reacts with ammonia, either within the exhaust or the ambient air to form ammonium chloride. Reaction of $PbBrCl$ with NH_4Cl gives rise to the observed double salts. The compound $3Pb_3(PO_4)_2{\cdot}PbBrCl$ is formed when phosphates are used as detergent additives in the fuel and is unlikely to be emitted in Britain where no such additives are used. $PbSO_4$ is a trace constituent formed when fuel sulphur is oxidized to sulphate.

It has long been known that vehicle-emitted lead undergoes fairly rapid chemical change when in the atmosphere, as particulate Br:Pb ratios change with residence time and season. It is, however, only recently that advances have been made in elucidating the mechanisms involved. Measurements in Britain have shown a range of lead compounds to be present in air at sites polluted only by vehicular lead [29, 30] – selected results appear in Table 2.9. The $PbBrCl$,

Table 2.9 Selected results of speciation study [30].

Site	Date	Compounds identified
Lancaster A6 road	17–19 May 1977	$PbSO_4{\cdot}(NH_4)_2SO_4$; $PbSO_4$
	30 Jan–6 Feb 1978	$PbSO_4{\cdot}(NH_4)_2SO_4$; α-$2PbBrCl{\cdot}NH_4Cl$; $PbBrCl{\cdot}3NH_4Cl$; NH_4Cl
	20 June–7 July 1978	$PbSO_4{\cdot}(NH_4)_2SO_4$; α-$2PbBrCl{\cdot}NH_4Cl$; $PbBrCl$
	1–4 Dec 1978	$PbSO_4{\cdot}(NH_4)_2SO_4$; α-$2PbBrCl{\cdot}NH_4Cl$; $PbBrCl$; $(NH_4)_2SO_4$
Shap M6 motorway	19–21 Sept 1978	$PbSO_4{\cdot}(NH_4)_2SO_4$
Red Scar, Preston M6 motorway	26–31 Oct 1977	$PbSO_4{\cdot}(NH_4)_2SO_4$
	14–21 Nov 1977	$PbSO_4{\cdot}(NH_4)_2SO_4$; $PbBrCl{\cdot}(NH_4)_2BrCl$
	21–28 Nov 1977	$PbSO_4{\cdot}(NH_4)_2SO_4$; $PbBrCl{\cdot}(NH_4)_2BrCl$
Preston A6 road	1–8 Aug 1977	$PbSO_4{\cdot}(NH_4)_2SO_4$; $PbBrCl$
	8–15 Aug 1977	$PbSO_4{\cdot}(NH_4)_2SO_4$
	4–11 Oct 1977	$PbSO_4{\cdot}(NH_4)_2SO_4$; $PbSO_4$
London A5 road	21–28 March 1978	$PbSO_4{\cdot}(NH_4)_2SO_4$
	4–11 Apr 1978	$PbSO_4{\cdot}(NH_4)_2SO_4$; $(NH_4)_2SO_4$

α-$2PbBrCl{\cdot}NH_4Cl$ and $PbBrCl{\cdot}2NH_4Cl$ probably arise directly from vehicle emissions. The latter two compounds were found only at urban sites, consistent with their reported emission primarily during city-type driving [5]. The predominant compound at all sites was $PbSO_4{\cdot}(NH_4)_2SO_4$, and this together with $PbBrCl{\cdot}(NH_4)_2BrCl$ is believed to arise from reaction of $PbBrCl$ with $(NH_4)_2SO_4$ subsequent to coagulation of primary vehicle-exhaust lead with the general aerosol, primarily neutral $(NH_4)_2SO_4$, and acid sulphates (NH_4HSO_4 and H_2SO_4) in the size range susceptible to rapid coagulation processes ($<0.1\ \mu m$).

$$2PbBrCl + 2(NH_4)_2SO_4 \rightarrow PbSO_4{\cdot}(NH_4)_2SO_4 + PbBrCl{\cdot}(NH_4)_2BrCl. \qquad (4)$$

The formation of lead sulphate is believed to arise from reaction with acid sulphates, which is accompanied by loss of gaseous HCl and HBr

$$6PbBrCl + 2NH_4HSO_4 \rightarrow 2PbSO_4 + 2PbBrCl{\cdot}NH_4Br + \alpha\text{-}2PbBrCl{\cdot}NH_4Cl + HCl + HBr \qquad (5)$$

$$PbBrCl + H_2SO_4 \rightarrow PbSO_4 + HCl + HBr \qquad (6)$$

The loss of HCl and HBr explains the observed decrease in the Br:Pb ratio in the atmosphere. The seasonal variation in which Br:Pb ratios are at a minimum in summer months is due to a higher ratio of acid: neutral sulphate at this time of year [30].

2.5.3 Organic lead

In Britain almost all of the tetraalkyllead petrol additive is used in the form of tetramethyllead or tetraethyllead with little usage of the mixed alkyl compounds, whilst in the US rather more of the mixed alkyls are in use. Tetramethyllead is more thermally stable, and generally less chemically reactive than tetraethyllead and European observations indicate that the former compound predominates in the atmosphere [17].

Both homogeneous and heterogeneous mechanisms for the removal of tetraalkyllead compounds from the atmosphere have been investigated [19]. Heterogeneous processes involve adsorption upon particles, with possible subsequent decomposition. These were found to be rather slow.

Homogeneous gas phase reactions with the reactive hydroxyl species (OH), triplet atomic oxygen (O^3P) and ozone (O_3) together with photolytic decomposition were found to be the most important breakdown mechanisms. These are typical reactions of hydrocarbon compounds in the atmosphere. The rates of the various removal processes are highly dependent upon environmental variables such as the intensity of sunlight and the concentrations of other pollutants which act as sources or sinks of the above reactive species. A summary of the breakdown rate predictions appears in Table 2.10.

Table 2.10 Estimated upper limit rates of tetramethyllead (TML) and tetraethyllead (TEL) decay in the middle of the day in a moderately polluted irradiated atmosphere [19].

	Concentration of reactive species		Decay rate (% h^{-1}) TML		Decay rate (% h^{-1}) TEL	
Decay path	s*	w*	s	w	s	w
OH attack	(1–3) × 10^{-7} ppm	(1–2) × 10^{-8} ppm	8–21	1–1.5	51–88	7–13
Photolysis	$z \sim 40°$	$z \sim 75°$	8	2	26	7
O_3 attack	100–200 ppb	40 ppb	1–2	0.5	9–17	4
$O(^3P)$ attack	10^{-8} ppm	10^{-9} ppm	<0.1	<<0.1	0.1	<0.1
Particulates	200 $\mu g\ m^{-3}$	200 $\mu g\ m^{-3}$	—	—	0.03	0.03
Total	—	—	16–29	3–4	67–93	17–23

*s, summer; w, winter

References

[1] US Environmental Protection Agency (1977), *Control Techniques for Lead Air Emissions*, EPA-450/2-77-012.

[2] Vandegrift, A. E., Shannon, L. J., Sallee, E. E., Gorman, P. G. and Park, W. R. (1971), Particulate Air Pollution in the United States, *J. Air Pollut. Control Assoc.*, **21**, 321–8.

[3] Lee, R. E. and von Lehmden, D. J. (1973), Trace Metal Pollution in the Environment, *J. Air Pollut. Control Assoc.*, **23**, 853–7.

[4] Central Unit of Environmental Pollution, Department of the Environment (1974), *Lead in the Environment and its Significance to Man*, HMSO, London.

[5] Hirschler, D. A. Gilbert, L. F., Lamb, F. W. and Niebylski, L. M. (1957), Particulate Lead Compounds in Automobile Exhaust Gas, *Ind. Eng. Chem.*, **49**, 1131–42.

[6] Habibi, K. (1973), Characterization of Particulate Matter in Vehicle Exhaust, *Environ. Sci. Technol.*, 7, 223–34.

[7] Chamberlain, A. C., Heard, M. J., Little, P. and Wiffen, R. D. (1979), The Dispersion of Lead from Motor Exhausts, Proceedings of the Royal Society Discussion Meeting, *Pathways of Pollutants in the Atmosphere*, London, 1977; *Phil. Trans. Roy. Soc. Lond. A.* **290**, 577–89.

[8] Yankel, A. J., von Lindern, I. H. and Walter, S. D. (1977), The Silver Valley Lead Study: The Relationship between Childhood Blood Lead Levels and Environmental Exposure, *J. Air Pollut. Control Assoc.*, **27**, 763–7.

[9] Daines, R. H., Motto, H., Chilko, D. M. (1970), Atmospheric Lead: Its Relationship to Traffic Volume and Proximity to Highways, *Environ. Sci. Technol.*, **4**, 318–22.

[10] Bevan, M. G., Colwill, D. M. and Hogbin, L. E. (1974), Transport and Road Research Laboratory Report 626.

[11] Little, P. and Wiffen, R. D. (1978), *Atmos. Environ.*, **12**, 1331–41.

[12] Waller, R. E., Commins, B. T. and Lawther, P. J. (1965), *Br. J. Ind. Med.*, **22**, 128–38.

[13] Harrison, R. M., Perry, R. and Slater, D. H. (1974), *Atmos. Environ.*, 8, 1187–94.

[14] Solomon, R. L., Hartford, J. W., Hudson, J. L., Neaderhouse, D. and Stuckel, J. J., Spatial Variation of Airborne Lead Concentration in an Urban Area, *J. Air Pollut. Control Assoc.*, **27**, 1095–9.

[15] Harrison, R. M. and Laxen, D. P. H. (1978), A Natural Source of Tetraalkyllead in Air, *Nature*, **275**, 238–40.

[16] Harrison, R. M. and Perry, R. (1977), The Analysis of Tetraalkyl Lead Compounds and their Significance as Urban Air Pollutants, *Atmos. Environ.*, **11**, 847–52.

[17] Harrison, R. M., Laxen, D. P. H. and Birch, J. (1979) Tetraalkyllead in Air: Sources Sinks and Concentrations, *International Conference – Management and Control of Heavy Metals in the Environment*, London, Sept., CEP Consultants Ltd., Edinburgh, pp. 257–61.

[18] Harrison, R. M. and Laxen, D. P. H. (1977), Organolead Compounds adsorbed upon Atmospheric Particulates: A Minor Component of Urban Air, *Atmos. Environ.*, **11**, 201–3.

[19] Harrison, R. M. and Laxen, D. P. H. (1978), Sink Processes for Tetraalkyllead Compounds in the Atmosphere, *Environ. Sci. Technol.*, **12**, 1384–92.

[20] Harrison, R. M., Laxen, D. P. H. and Birch, J. (1980), A Specific Method for 24–48 Hour Analysis of Tetraalkyllead in Air, *Sci. Tot. Environ.*, **14**, 31–42.

[21] Georgii, H. W. and Jost, D. (1971), *Atmos. Environ.*, **5**, 725–7.

[22] Brief, R. S., Jones, A. R. and Yoder, J. D. (1960), *J. Air Pollut. Control Assoc.*, **10**, 384–8.

[23] Bové, J. L. and Siebenberg, S. (1970), *Science*, **167**, 986–7.

[24] Colucci, J. M., Begeman, C. R. and Kumler, K. (1968), Paper at 61st Annual Meeting APCA, St. Paul, Minn.

[25] Martens, C. S., Wesolowski, J. J., Kaifer, R. and John, W. (1973), *Atmos. Environ.*, 7, 905–15.

[26] Perry, R., Whitelaw, K. and Harrison, R. M. (1978), A Correlation Study of Vehicle-Generated Air Pollutants, *Water Air Soil Pollut.*, **10**, 115–127.

[27] Laxen, D. P. H. and Harrison, R. M. (1977), The Highway as a Source of Water Pollution: An Appraisal with the Heavy Metal Lead, *Water Res.*, **11**, 1–11.

[28] Smith, W. H. (1976), Lead Contamination of the Roadside Ecosystem, *J. Air Pollut. Control Assoc.*, **26**, 753–66.

[29] Biggins, P. D. E. and Harrison, R. M. (1978), Identification of Lead Compounds in Urban Air, *Nature*, **272**, 531–2.

[30] Biggins, P. D. E. and Harrison, R. M. (1979), The Atmospheric Chemistry of Automotive Lead, *Environ. Sci. Technol.*, **13**, 558–65.

3

Lead in water

3.1 Introduction

Elevated levels of lead in water arise principally from industrial discharges, highway runoff and weathering processes in areas of natural lead mineralization. Once the lead has entered a water body its mobility and distribution will be closely controlled by its chemical forms. These will alter according to the nature of the particular water body.

During its transport by water, the lead may interact detrimentally with the aquatic life. Furthermore, it will become available for abstraction into public water supplies. In both cases the precise behaviour and effects will be determined by the chemistry of the lead.

3.2 Sources of lead in surface waters

Lead is deposited from the atmosphere directly onto water surfaces. It is also contributed by effluents arising from industrial, domestic and transport activities, as well as being derived from natural sources, in particular areas of lead mineralization.

Direct deposition, either by dry or wet pathways (Section 4.2) is only significant in the case of larger water bodies such as lakes or oceans. It has been estimated that more lead now enters the North Sea *via* atmospheric input, 15 000 tonne y^{-1}, than is contained in the known water-borne inputs of 3600 tonne y^{-1} [1]. The estimated magnitude of different pathways of lead input to all the worlds oceans, including both industrial and natural sources is shown in Table 3.1.

The remaining sources of lead are in the form of liquid effluents. These may be usefully considered in terms of point sources – readily identified large discharges – and as diffuse or non-point sources, which are smaller in scale and more widespread, and hence harder to quantify. A rigid distinction should not, however, be drawn between these two types of source.

Table 3.1 Approximate lead input to all oceans, [2].

Input	(Tonnes yr^{-1})
Industrial Inputs	
Aerosols (gasoline)	37 000
Aerosols (smelters and forest fires)	3 000
Rivers and sewers (soluble, mainly from aerosols)	60 000
Rivers and sewers (solids)	200 000
Pre-industrial inputs	
Aerosols	1 000
Rivers (soluble)	13 000
Rivers (solids)	100 000

3.2.1 Point sources

3.2.1.1 Industrial effluents

Systematic data on lead discharges to water from industrial sources are not readily available. It is thus only possible to indicate the order of magnitude of these discharges. It is reported from England that a medium-sized multi-product dyestuff company produces an effluent with a lead loading of typically 13 kg(Pb) day^{-1}, although future discharges to a sewage treatment works must be limited to <1.7 kg(Pb) day^{-1} (Section 6.4.1.5a) [3]. Similarly a zinc–lead smelter in Australia discharged typically 11 kg(Pb) day^{-1} into Cockle Creek in 1975, although stricter limits require this discharge to be reduced (Section 6.4.1.5d) [4]. In contrast the South Teeside steelworks discharged 310 kg(Pb) day^{-1} of lead in 1972 [5], but in this case the discharge was to an estuary for which discharge limits are usually higher. Nonetheless, this discharge has probably been reduced by now as a result of improved treatment [5].

A detailed analysis of the various effluent streams contributing to the final discharge from the South Teeside steelworks is presented in Table 3.2 (Fig. 3.1). The discrepancy between the summed source total and that measured in the final discharge probably reflects the inadequacy of performing a mass balance using data obtained as infrequent spot samples and showing a considerable range of concentrations (see Table 3.9 on p. 40).

3.2.1.2 Sewage effluent

Sewage treatment plants receive domestic wastewaters and in some instances industrial effluents and/or storm water runoff, all of which will be contaminated to some extent with lead. Treatment can significantly reduce the lead content of the effluent (Section 6.4.2), but some lead is still discharged with the final effluent [6] (Table 3.3).

It has been estimated that the daily *per capita* contribution of lead from

Table 3.2 Lead loading in various effluents in the South Teeside Steelworks, [5].

Source	Loading (kg day^{-1}) Dissolved	Suspended solids
Holme Beck	2.0	0.4
Ferro-manganese Plant	2.7	18.9
Blast Furnaces	6.5*	122*
BOS Plant	3.8	49.5†
Lackenby Works	0.7‡	2.3‡
Cleveland Coke Ovens	§	§
Electric Arc Plant	§	47
Total (corrected)¶	16	157
Final effluent (measured)	11	300

*Includes South Bank coke ovens.

†Loading after first settling pond = 25 kg day^{-1}.

‡Lackenby works plus Kinkerdale Beck.

§No data.

¶Corrected to take account of settling of suspended solids.

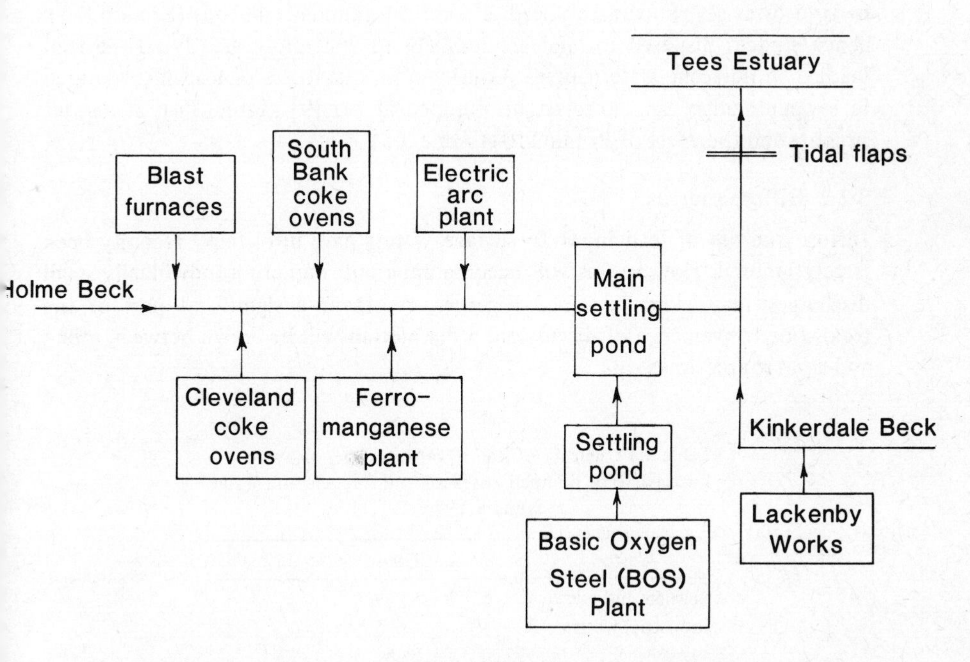

Fig. 3.1 Effluent sources and network of effluent flows in South Teeside Steelworks (based on [5]).

Table 3.3 Quantities of lead present in the sewage liquors and sludges, and final effluent of the Oxford Sewage Treatment Works over 24 h [6].

Sampling occasion	Lead flow (kg day^{-1})				
	Raw sewage	Primary sludge	Settled sewage	Activated sludge	Final effluent
1	3.66	2.54	0.880	0.600	0.410
2	3.02	1.67	0.529	0.559	0.000
3	1.47	1.47	0.450	0.490	0.014
4	1.25	1.24	0.426	0.400	0.246

secondary sewage effluent to a watershed in Virginia, USA is 0.0026g [7]. Similar estimates can be derived from the data for the Oxford sewage treatment plant (Table 3.3), whilst the mean daily *per capita* lead discharge from 5 treatment plants serving 1 million people in Sweden is *ca.* 0.01 g [8]. These data imply a lead discharge of up to 100 kg day^{-1} from sewage treatment plants serving a total population of 10 million.

Whilst considering sewage effluent, it should be remembered that the sludges separated during treatment contain large amounts of lead (Table 3.3). These sludges can ultimately act as a source of lead in water, following their dispersal to land or at sea. Fortunately lead is relatively immobile in soils (Section 4.5), hence sludges disposed to land are unlikely to contribute greatly to the lead loading in adjacent watercourses. Considerable quantities of lead incorporated in sewage sludge can, however, be dumped at sea [9] (Table 3.4), at various sites around the coast of Britain [10] (Table 3.5).

3.2.2. Diffuse sources

Diffuse sources of lead input to surface waters have until fairly recently been largely ignored. However, as will become apparent, numerous individually small discharges may, when summed together, represent a significant part of the total. For convenience of discussion, a distinction will be drawn between urban and rural source areas.

Table 3.4 Quantity of lead in various types of wastes licensed to be dumped at sea around England and Wales. January 1976, [9].

Waste	Quantity (kg day^{-1})*
Sewage sludges	537
Industrial wastes	5.5
Fly ash	55
Dredgings†	8750

*Derived from annual limits.

†Incomplete data. In many cases this is probably a recycling process within the marine environment and does not represent new inputs.

Table 3.5 Inputs of lead to the marine environment from sewage sludge dumping, [10].

Dumping ground	Type of sludge*	Input (kg day^{-1})†
Barrow Deep, Thames Estuary	D	274
Liverpool Bay	P, S, D	82
Bristol Channel	P, S, D	36
Spurn Head off River Humber	P, S, D	74
Off Harwich	P, S, D	11
Nab Tower, Isle of Wight	P, S	5
Off Plymouth	P, S, D	3
Off Exeter	D	5
Off River Tyne	P	Not operational in 1977
Off Belfast Lough	P, S	10
Garroch Head	P, S, D	118
Off Forth Estuary	P	Not operational in 1977
Total		618

*P = primary, S = secondary and D = digested. †Derived from annual inputs.

3.2.2.1 Urban areas

Substantial quantities of lead emitted in vehicle exhaust gases are deposited directly onto the road surface. Shaheen [11] has carried out one of the most comprehensive studies of this particular source of lead, based on the Washington DC Metropolitan Area. The average deposition rate for lead was 7.9×10^{-3} g per axle-km. However, an important finding of this study was that the accumulation rate of materials deposited on road surfaces prior to washoff is not linearly related to time. The accumulation levels off after several days, due to removal of the deposited material by the passage of vehicles. Thus the ratio of lead accumulated on the road surface after 3 days to that accumulated after 1 day was found to be only 1.2 : 1.

The highway is undoubtedly the most important source of lead in urban runoff [12, 13]. Rain washes the lead, most of which is associated with particulate matter, off the road, transporting it *via* an efficient drainage system directly to local water courses or in the case of a combined sewerage system to a treatment plant. It has been found that a simple exponential model can be used to describe the removal of surface contaminants [12]

$$L_i - L_r = L_i(1 - e^{-kR})$$

where L_i and L_r are the surface loading (mg m^{-2}) initially and after time t, R is the cumulative rainfall (mm) (actual or effective) up to time t and k is the removal constant (mm^{-1}). The removal constant is high for soluble surface contaminants, ($k \simeq 2$ mm^{-1}), and considerably lower, ($k \simeq 0.1$ mm^{-1}) for suspended sediments, i.e. particulate lead. In the former case, a 90% removal of surface contaminant occurs after only 2–3 mm of rainfall, this must increase to >20 mm of rainfall for a 90% removal of suspended solids [12].

Hedley and Lockley [14] have examined the runoff from an 800 m section of

7-lane urban motorway. The average daily runoff of lead was 190 g day^{-1}, the level being much higher in winter than in summer. A small percentage of this runoff, 1–4%, was due to lead contamination of the de-icing salt used extensively during the winter.

The total amount of lead discharged in an urban area averaged over time, will depend upon the percentage of paved area, the traffic density and factors such as street cleaning. Bryan [15] has calculated the quantity of lead discharged in waters draining urban catchments in Durham, NC (Table 3.6). The estimated mean daily discharge of 4–8 g ha^{-1} day^{-1} is probably a slight overestimate in comparison with the results of the more detailed studies cited below. However, it allows the calculation of a mean daily *per capita* lead discharge of 0.2–0.3 g *per capita* day^{-1} which can be usefully compared with the much lower values of 0.001–0.01 g *per capita* day^{-1} for lead discharged with treated sewage effluent (Section 3.2.1.2).

Table 3.6 Lead discharge from 3 drainage basins in Durham, NC [15].

	Basins		
	N	W	N+W
Area (ha)	281	152	433
Population density (per ha)	26	20	24
Land use (% of basin area)			
Residential	49	79	60
Commercial/Industrial	28	2	19
Other	23	19	21
Paved streets and parking lots (% of basin area)	24	13.4	20.2
Yield of lead			
kg ha^{-1} y^{-1}	3.0	1.5	2.6
g ha^{-1} day^{-1}	8.2	4.1	7.1
g *per capita* day^{-1}	0.32	0.21	0.30

A more detailed study of sub-catchments in a Virginia watershed [7] yielded lead discharge rates of 0.92 g ha^{-1} day^{-1} for a low density residential area and 2.3 g ha^{-1} day^{-1} for drainage from a commercial area. Similar results have been obtained in England for a small catchment of mature housing, where the discharge was 0.11 g ha^{-1} day^{-1} [13]. Traffic flow in the area was reported as 9500 vehicles day^{-1}.

It is important to remember that the lead discharged with storm watei runoff provides an intermittent and highly variable loading which will depend upon the characteristics of the rainfall event, the time since the last rainfall and the efficiency with which the lead is removed [12]. Careful sampling programmes must therefore be designed in order to quantify adequately the discharge characteristics of the fast response urban drainage network.

3.2.2.2 Rural areas

Highway drainage waters will still contribute to lead in rural watercourses. The degree of this contribution, however, will depend upon the road density and the extent of vehicular activity. Nevertheless, the predominant source of water-borne lead in rural areas will usually be the weathering of parent rocks, together with the dissolution and washoff of lead in soils, as well as that deposited on vegetation. The former source will be greatest in areas of lead mineralization, particularly when mining operations render the material more available for solution processes.

Airborne lead deposited on a natural watershed will contribute only slightly to the lead in the runoff waters. Andren and co-workers [16] have studied the lead budget for the Walker Branch Watershed in Oak Ridge, Tennessee. The data are summarized in Table 3.7 and show that only 2–3% of the lead input is transferred out of the watershed by the stream. The stream discharge of lead is equivalent to ~ 0.016 g ha^{-1} day^{-1}, one to two orders of magnitude less than the discharge of lead from an urban catchment.

Table 3.7 Cycling of lead through the Walker Branch Watershed, Jan.–June 1974, (based on [16]).

	Lead movement (g ha^{-1})
Input	
Wetfall	115
Dry deposition	<56.5
Total	115–171
Output	
Dissolved	2.66
Suspended	0.21
Total	2.9
Percentage of input	2–3%

3.3 Concentrations of lead in water

Concentrations of lead in water are presented either in terms of total metal or separately as soluble and particulate metal. The latter distinction is arbitrarily based upon sample filtration, usually with a 0.45 μm membrane filter. As a general rule, it can be assumed that the higher the total lead concentration, the higher will be the ratio of particulate to soluble lead. The reasons for this will become apparent during the discussion of the chemistry of lead in water (Section 3.4).

The particulate lead measured in a sample of water is that associated with the suspended sediment. However, prior to suspension, or following settlement of the suspended matter, the particulate lead will be associated with the less mobile bed sediment. Indeed, much of the lead entering quieter waters, i.e. lakes and oceans, will become incorporated in the bed or bottom sediment. It must be realised however, that the water bodies are rarely at equilibrium and chemical changes can result in the release of lead from the bed sediments into the overlying waters (Section 3.4).

3.3.1 Source waters

The concentrations of lead in polluted source waters lie typically in the range 1-100 mg dm^{-3} (Table 3.8). These concentrations are, however, frequently reduced during treatment, prior to discharge to the receiving waters. Prater [5] has presented a detailed breakdown of the lead concentrations in different effluent streams in the South Teeside steelworks (Table 3.9, see also Fig. 3.1). Comparison of these data with the daily loadings previously presented in Table 3.2 illustrates the important point that the highest concentrations are not necessarily associated with the most significant pollution source. In this example,

Table 3.8 Reported lead levels in industrial waste waters before treatment.

Industry	Lead (mg dm^{-3})	Reference
Battery manufacture		
Particulate	5–48	[17]
Soluble	0.5–25	
Plating	2–140	[17]
Plating	0–30	[17]
Television tube manufacture	400	[17]
Mine drainage	0.02–2.5	[17]
Mine process water	0.018–0.098	[17]
Tetraethyllead manufacture		
Organic	127–145	[17]
Inorganic	66–85	
Tetraethyllead manufacture	45	[17]
Multi-product dyestuffs	3	[3]
Steelworks		
Soluble	<0.05–9.5	[5]
Particulate	0.016–49	

Table 3.9 Lead concentrations in various effluents in the South Teeside Steelworks, [5].

Source	Lead (mg dm^{-3}) Dissolved		Suspended Solids	
	Mean	Range	Mean	Range
Holme Beck	<0.09	<0.05–0.32	0.019	0.004–0.07
Ferro-manganese Plant	≤0.06	<0.05–0.08	0.42	0.016–0.87
Blast Furnaces	≤0.08	<0.05–0.20	1.5	0.07–2.3
BOS Plant	≤1.7	<0.05–9.5	22	0.14–49
Lackenby Works	<0.05	<0.05	0.17	0.026–0.54
Cleveland Coke Ovens	–*	–	–	–
Electric Arc Plant	–	–	15	–
Final Effluent (measured)	≤0.07	<0.05–0.15	1.9	0.3–3.7

*Indicates no data.

the Basic Oxygen Steel (BOS) plant has the highest lead concentration, yet the Blast Furnaces, with a concentration an order of magnitude lower, represent the major source of lead in the final effluent.

Lead concentrations typical of those found in raw and treated sewage effluent are indicated by the data for the Oxford Sewage Treatment Works (Table 3.10) [6]. The concentrations are highly variable, especially in the final effluent. In this particular case, the final effluent concentrations are only just above background levels in the receiving water.

Table 3.10 Concentrations of lead in sewage and final effluent of the Oxford Sewage Treatment Works, 24 h values, [6].

Sampling occasion	Lead concentration (mg dm^{-3})		
	Raw sewage	Settled sewage	Final effluent
1	0.250	0.060	0.028
2	0.200	0.034	<0.001
3	0.100	0.031	0.001
4	0.081	0.020	0.016

Lead concentrations in storm water runoff are highly variable, being dependent, amongst other factors, upon the level and type of activity in a particular area (Table 3.11). Definition of typical lead concentrations is nevertheless virtually impossible, as they vary considerably during a storm, generally showing a decline as the lead is washed away (Fig. 3.2). The ultimate concentration will be related to that in the rain causing the runoff. This will typically be in the range 0.1–0.4 mg dm^{-3} in urban areas [19]. However, whilst the lead in rain is mostly soluble, that found in the runoff waters occurs largely as insoluble particulate lead, due to adsorption and/or chemical precipitation processes (Section 3.4).

3.3.2 Receiving waters

Concentrations of lead in natural waters are generally very low, 0.001–0.003 mg dm^{-3}, rising in waters draining areas of lead mineralization up to 0.05–0.02

Table 3.11 Average peak concentrations of lead in storm runoff waters, [18].

Land cover type	% Impervious area	Average peak lead concentration (mg dm^{-3})
Low activity rural	2.7	<0.1
High activity rural	5.1	<0.1
Low activity residential	16	0.6
High activity residential	32	2.1
Low activity commercial	12	0.2
High activity commercial	35	1.7
Central business district	80	0.9

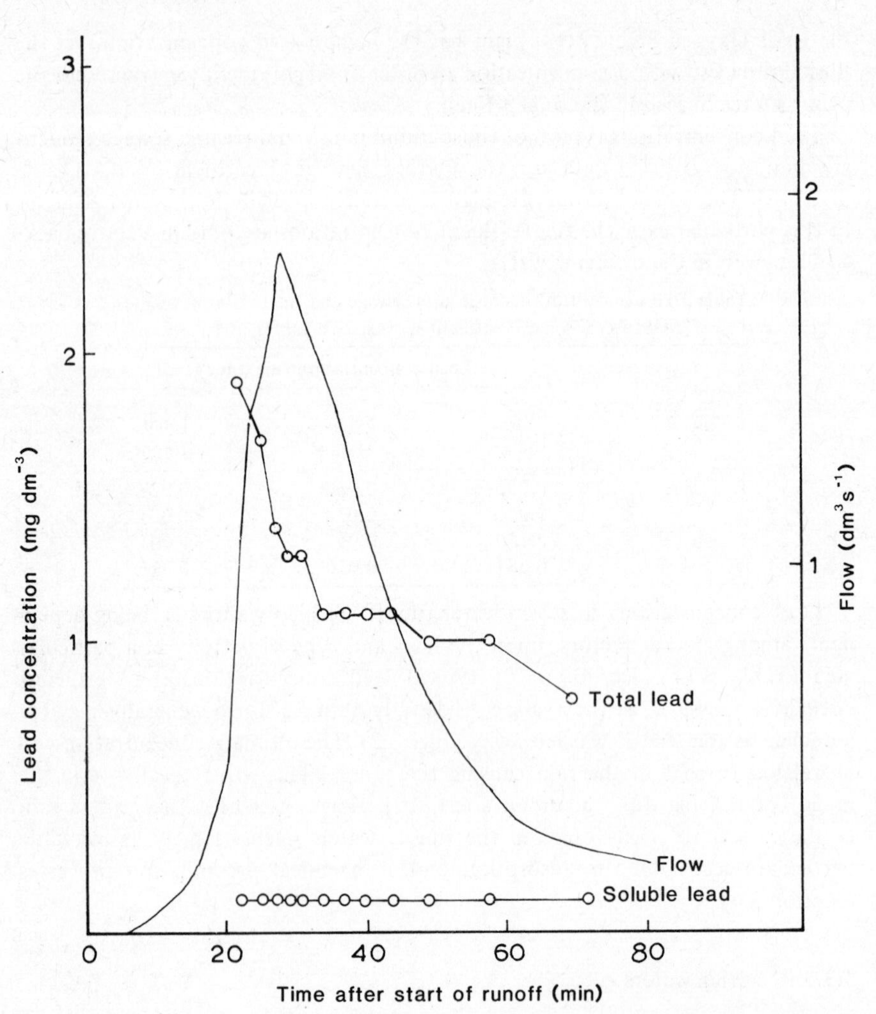

Fig. 3.2 Concentration of lead in storm water runoff from a highway [11].

mg dm^{-3} (Table 3.12) [20, 21]. Waters receiving polluted discharges will show elevated concentrations dependent upon the quantity of lead discharged and the dilution capacity of the receiving water. Highly variable concentrations will result within waters receiving intermittent storm water runoff.

It is a common misconception that the concentration of lead in a particular discharge will be representative of its significance as a source of pollution. It is thus worth re-emphasizing (see also Section 3.3.1) that the effect on a receiving water will depend upon the mass discharge rate of the lead (mg s^{-1}), which is a function of both concentration (mg dm^{-3}) and the flow rate of the discharge (dm^3 s^{-1}), i.e. mass discharge rate (mg s^{-1}) = concentration (mg dm^{-3}) $\times$ flow

Table 3.12 Lead concentrations in waters draining mineralized and unmineralized catchments in south-west England, [20].

Catchment	Lead concentration ($mg\ dm^{-3}$)	
	Mean	Range
Mineralized		
A	0.003	0.002–0.004
B	0.007	0.003–0.015
C	0.073	0.011–0.200
Unmineralized		
A	<0.001	<0.001–0.0015
B	0.002	<0.001–0.003
C	0.003	0.002–0.008

rate ($dm^3\ s^{-1}$). The concentration increase within the receiving water will thus depend upon its own rate of flow ($dm^3\ s^{-1}$) at the time of discharge and the mass discharge rate of lead ($mg\ s^{-1}$) into it, i.e. concentration increase ($mg\ dm^{-3}$) = mass discharge rate ($mg\ s^{-1}$) ÷ receiving water flow rate ($dm^3\ s^{-1}$).

Lead concentrations in ocean waters are generally lower than those of inland waters. In deep ocean waters, concentrations fall as low as 0.000 004 $mg\ dm^{-3}$ [22], rising in coastal waters which receive surface runoff to values in excess of 0.005 $mg\ dm^{-3}$ [21]. Concentrations of lead measured in the coastal waters off the west coast of England and Wales are shown in Fig. 3.3. They reflect the impact of runoff from mineralized areas, as well as that due to industrial and domestic discharges. It should be noted that the concentrations are similar in the areas receiving runoff from both industrial and natural sources of lead.

3.3.3 Ground water

Water is constantly infiltrating through soils into the underlying bedrock, where it circulates as ground water, before generally resurfacing at some later time to provide the baseflow in streams and rivers. It has already been noted that lead is effectively immobilized in the top layers of soils (Section 4.5), and hence little will be carried down into the ground water. Concentrations of lead measured in ground water are generally low, the majority of waters having concentrations of less than 0.01 $mg\ dm^{-3}$ [23]. In this respect the concentrations are similar to those of surface waters. Elevated levels are found only in areas with water of low pH or high chloride content and with relatively high temperatures [23]. Concentrations can also be expected to be enhanced locally due to leaching from land-fill sites.

3.3.4 Sediments

Sediments, particularly those on the ocean bed, represent the final resting place (on a human, not geological time scale) for much of the lead dispersed through

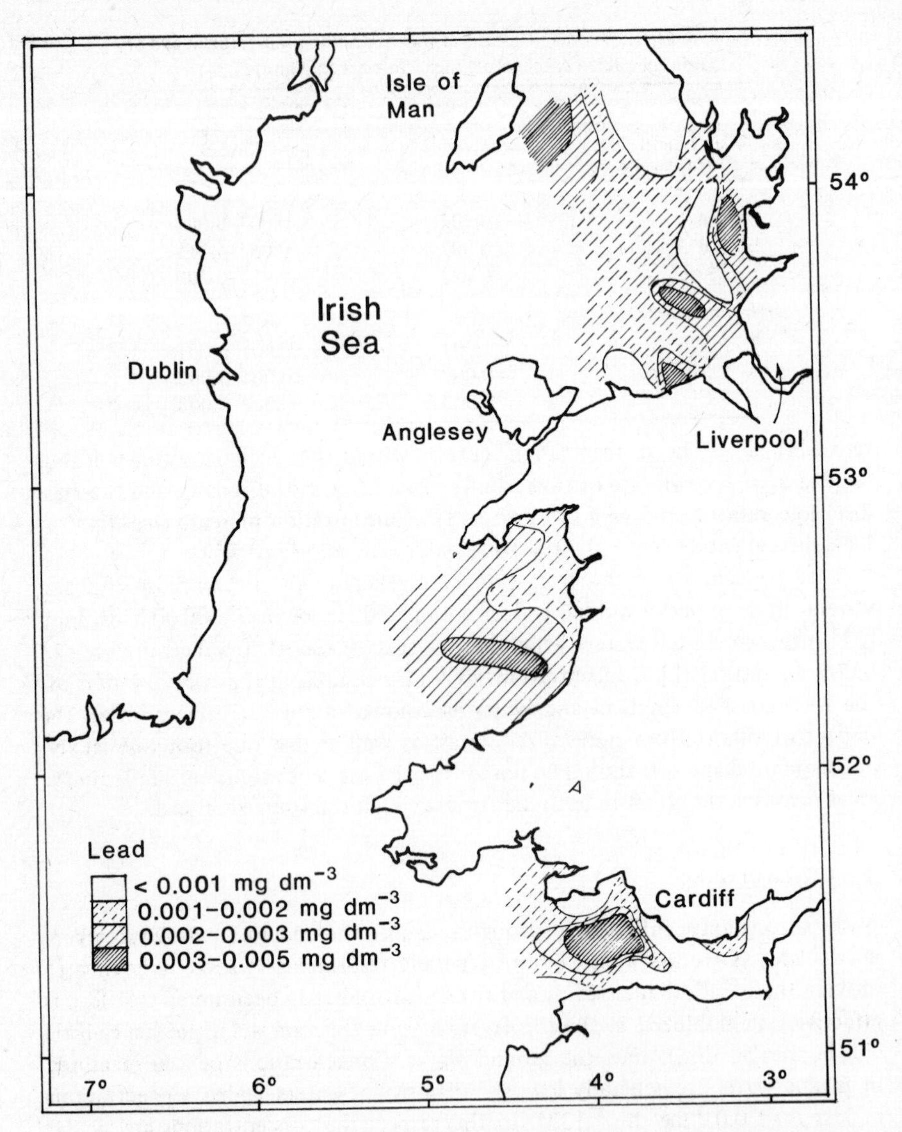

Fig. 3.3 Distribution of lead concentrations in the eastern coastal waters of the Irish Sea [21].

the environment. Indeed, profiles of lead concentration in lake sediments, which show a dramatic increase towards the surface, i.e. in the more recently deposited sediments, have been used to highlight the enhanced emissions of lead to the environment over the last century.

Concentrations of lead associated with the particulate matter in contaminated discharges are summarized in Table 3.13. Concentrations within the receiving water will be reduced by mixing with less contaminated particulate matter. At the same time, lead may be released from the solids, or on the other hand, further lead may become fixed to the particulate matter (Section 3.4). The distribution of concentrations in stream bed sediments in England and Wales, based on *ca.* 50 000 samples, is shown in Fig. 3.4. Concentrations in deep ocean bottom sediments (clays) have been found to range from 50 to 150 mg kg^{-1}, decreasing towards coastal areas to *ca.* 20 mg kg^{-1} [23], except in those areas receiving contaminated land based discharges, where concentrations may rise to over 400 mg kg^{-1} [21]. The increased concentration in deep ocean sediments is thought to be due to a migration of sediment away from coastal areas.

Table 3.13 Lead concentrations of the solids entering surface waters.

Source	Lead concentration (mg kg^{-1} of solid)	Reference
Street dust – urban	1 000–4 000	[12]
Street dust – rural	440	[12]
Suspended sediment in highway runoff	5 800	[12]
Suspended sediment in highway runoff	3 100	[12]
Settleable solids in highway runoff	16 000	[12]
Sewage sludge for sea disposal	100–1400	[10]
Suspended sediment in mineralized area	1 000–8 000	[20]

Aston and Thornton [20] have suggested that the concentration of lead in stream sediments of <200 μm can be used to indicate the maximum likely lead concentration in the overlying stream water. Based on their measurements, they tentatively associate a sediment lead concentration of 500 mg kg^{-1} with a maximum lead concentration in the overlying water of 0.05 mg dm^{-3}.

3.3.5 Lead in drinking water

An important source of lead leading to human exposure is that contained in drinking water. Treatment processes for public water supplies are unlikely to reduce the concentration of filterable lead in the raw water, except where processes such as precipitation water softening are used [25]. Fortunately, however, the concentrations of lead in the raw water are generally fairly low, <0.01 mg dm^{-3}. Many tap water samples, nevertheless, reveal significantly elevated lead levels, which may occasionally rise to over 1 mg dm^{-3}. These elevated concentrations are found in areas where the plumbing system is based on lead pipes.

The highest lead concentrations occur in standing water first drawn from the tap. They then generally decline after a few litres of water have been drawn off. The measured concentrations of lead in tap water must be viewed in the light of the national and international standards, which have been variously set at 0.05 or 0.1 mg dm^{-3} [25]. The preliminary results of a recent survey of tap water in England, Wales and Scotland suggest that 1.7 million households (*ca.* 9.2%) have a first draw lead concentration in excess of even the higher standard. This number falls to 0.8 million households (*ca.* 4.4%) when daytime samples are considered [25].

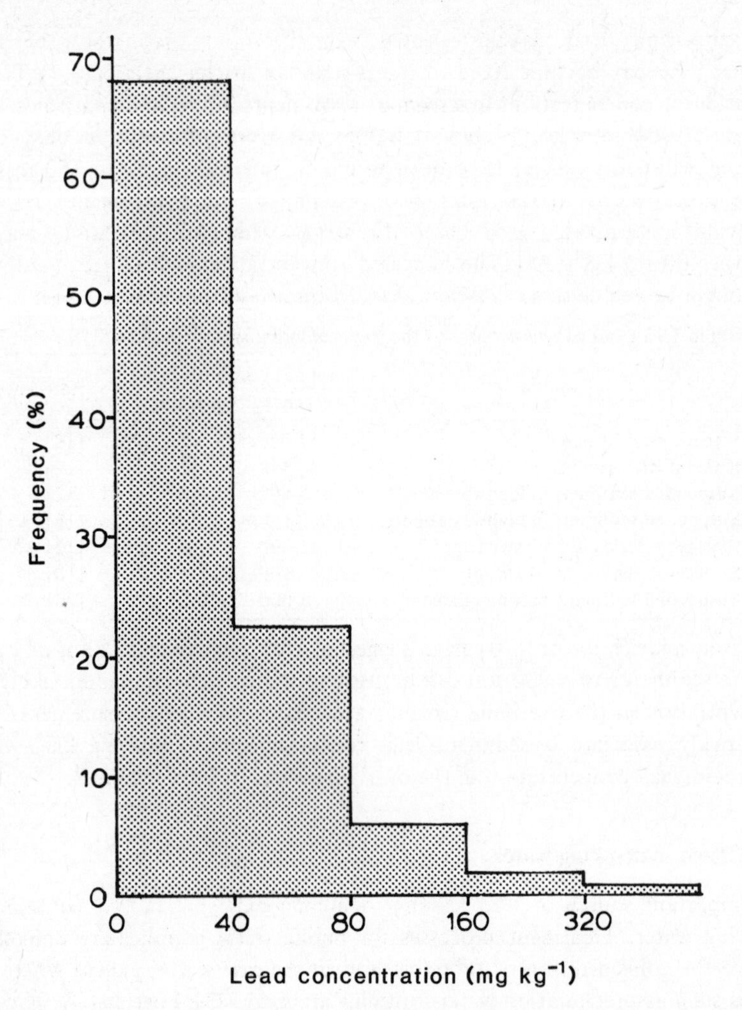

Fig. 3.4 Distribution of lead concentrations in stream bed sediments of England and Wales [24].

3.4 Chemistry of lead in water

The physico-chemical forms in which lead is found in water will exert a powerful control on its dispersal from the point of discharge, as well as on its subsequent mobility and ultimate distribution in different receiving waters. The form of the lead prior to its discharge will also play a part in establishing the effectiveness of particular effluent control strategies (Chapter 6). Once in the receiving water the toxic effects of the lead to aquatic life will be intimately related to the chemical form(s). However, despite its importance, the chemistry of lead in water is still only poorly understood, although increasing attention is now being paid to the subject [12].

Studies of the various physico-chemical forms of lead and their interactions in different types of water are hampered by the wide range of their possible forms and the extreme difficulty in distinguishing them. A summary of possible forms is given in Fig. 3.5, based on the classification of Stumm and Bilinski [26], which uses size association as the distinguishing feature. It should be immediately apparent that the use of a 0.45 μm filter to distinguish soluble and particulate lead is merely an arbitrary distinction. A large part of the lead passing through a 0.45 μm filter may be associated with colloidal material, with, in some circumstances, only a small fraction being truly soluble.

It is only possible, in a short space, to outline the importance of the different forms of lead and to look at the principal controls that determine which form(s) might exist in a particular water. It is, furthermore, necessary to draw largely on the more theoretical evidence, as there is a lack of data on the forms of lead in polluted waters, and only sparse information, largely contradictory, relating to more natural waters [12].

3.4.1 Solubility control

Typically three lead compounds limit the solubility of lead in water, via the following equilibria

$$Pb(OH)_2(s) \rightleftharpoons Pb^{2+} + 2OH^- \; (K_{s_1} = 10^{-15.5} \text{ at } 25° \text{ C})$$

$$PbCO_3(s) \rightleftharpoons Pb^{2+} + CO_3^{2-} \; (K_{s_2} = 10^{-13.2} \text{ at } 25° \text{ C})$$

$$Pb_3(OH)_2(CO_3)_2(s) \rightleftharpoons 3Pb^{2+} + 2OH^- + 2CO_3^{2-} \; (K_{s_3} = 10^{-46.8} \text{ at } 25° \text{ C})$$

(K_s values from [23])

Lead ions (Pb^{2+}), however, have a strong tendency to form ion pairs, principally $PbHCO_3^+$ and $PbCO_3^0$ at the pH of most waters. The formation of these species reduces the Pb^{2+} concentration and drives the above equilibria to the right, enhancing the lead solubility. A similar enhancement of lead solubility occurs when organic compounds complex with the Pb^{2+} ions. The calculated equilibrium solubility of lead, allowing for the formation of inorganic ion pairs, is illustrated as a function of pH in Fig. 3.6.

The identification of lead in association with carbonates in the sediment phase of natural waters implies that lead solubility equilibria can indeed determine the particulate lead species in certain waters. Precipitated lead salts will almost certainly be present in the particulate phase of effluents treated by pH control (Section 6.4.1). Clearly, if lead is present in waters as carbonate/hydroxy precipitates, then any reduction in the water pH will increase the concentration of soluble lead. In most natural waters, however, the concentration of soluble lead is less than that predicted by equilibrium solubility considerations and other mechanisms will account for the presence of particulate lead.

Size	← 1nm		— 10nm	— 100nm	— 1000nm		
	Soluble			Colloidal		Particulate	
Metal species	Free metal ions	Inorganic ion pairs: Organic chelates	Organic complexes	Metal species bound to high molecular weight organic material	Metal species adsorbed on colloids	Metals incorporated with organic particles and remains of living organisms	Mineral solids: Metals adsorbed on solids: Precipitates and co-precipitates
Example	Pb^{2+}	$PbHCO_3^+$ Pb–EDTA	Pb–fulvic acid	Pb–humic acid	$Pb–Fe(OH)_3$ $Pb–MnO_2$	Pb–organic solids	Pb–clay $PbCO_{3(s)}$

Fig. 3.5 Lead species in water (based on [26]).

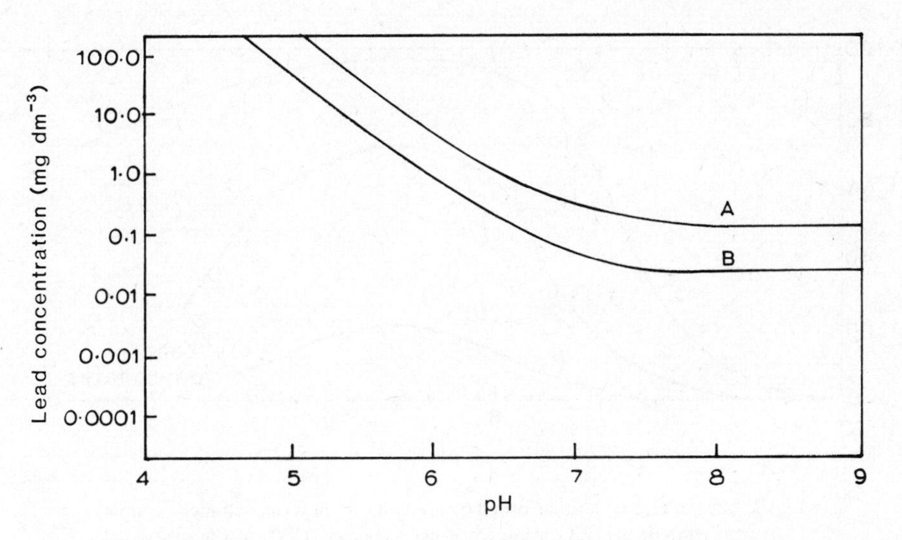

Fig. 3.6 Equilibrium lead solubility as a function of pH and total carbonate concentration (based on 27].

A. $[Na^+] = 10^{-4}$ $[Ca^{2+}] = 10^{-4}$ $[CO_3]_T = 10^{-3.5}$

B. $[Na^+] = 10^{-3}$ $[Ca^{2+}] = 10^{-3}$ $[CO_3]_T = 10^{-2.7}$

All concentrations in mol dm^{-3}.

3.4.2 Suspended sediments

Both specific adsorption and ion exchange mechanisms may operate to remove lead from solution onto suspended or even settled sediments. The fine fraction of waterborne sediments can show a considerable cation exchange capacity. Hem [27] has applied a theoretical model to the ion exchange behaviour of lead in natural waters, and finds that the proportion of lead associated with sediment depends upon the pH and the cation exchange capacity of the sediment (Fig. 3.7). The cation exchange capacities shown in Fig. 3.7 are probably typical of those for a suspended sediment concentration of somewhere between 1 and 100 mg dm^{-3}. These results suggest that a significant proportion of lead in water may be reversibly adsorbed on suspended sediments.

Clearly an ion exchange control on the chemistry of lead is dependent upon the pH of the water, and changes will thus occur with a change in pH. This process is also susceptible to changes in the concentration of the competing ions such as Na^+ and Ca^{2+}. The model developed by Hem predicts that in estuarine waters most of the lead adsorbed by ion exchange would be brought into solution.

3.4.3 Organic matter

Lead may interact in several ways with organic substances in water (Fig. 3.5).

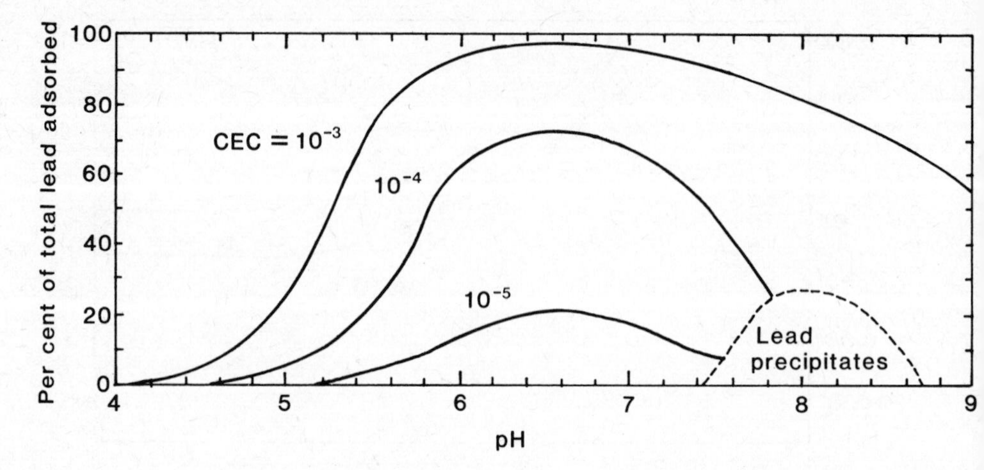

Fig. 3.7 Percentage of lead adsorbed or precipitated in water-sediment systems, as a function of pH and Cation-Exchange Capacity (CEC) of the suspended sediments [27].
$[Na^+] = 10^{-3}$ $[Ca^{2+}] = 10^{-3}$ [Total Lead] $= 10^{-7}$ CEC $= 10^{-5}$ to 10^{-3}.
All concentrations in mol dm^{-3}.

Sewage effluents are particularly rich in organic matter and they may contain significant quantities of the chelating reagent ethene diamine tetraacetic acid (EDTA) as well as a number of amino acids, all of which are capable of forming strong complexes with lead. In surface waters, however, the major organic compounds are the ill-defined humic and fulvic acids, which may also form complexes with lead. The formation of such complexes is strongly pH dependent and shows a reduction as the pH is lowered. Competition with other cations in solution will also affect any lead–organic matter interactions. This will occur during the transition from fresh water to sea water with its far higher concentration of competing cations (Ca^{2+}, Mg^{2+}, Na^+) and ligands (Cl^-, $SO_4{}^{2-}$, $CO_3{}^{2-}$). Actual evidence regarding the existence and magnitude of lead–organic interactions in surface waters is, nevertheless, still very much in dispute.

3.4.4 Colloidal hydrous ferric and manganese oxides

Lee [28] has argued persuasively that scavenging by colloidal hydrous ferric and manganese oxides ($Fe(OH)_3$ and MnO_2) may well act as the most significant sink for heavy metals in aerobic waters. Lead shows a particularly strong affinity for these hydrous oxides, the order of adsorption being $Pb^{2+} > Cd^{2+} > Zn^{2+} > Ca^{2+} > K^+$ [12]. The sorption process is rapid and may well be largely irreversible, particularly if the metal is incorporated with the hydrous oxide as it is precipitated. The adsorption is, however, strongly pH dependent, increasing with an increase in pH (Fig. 3.8) [29].

Hydrous ferric and manganese oxides are readily reduced and hence become

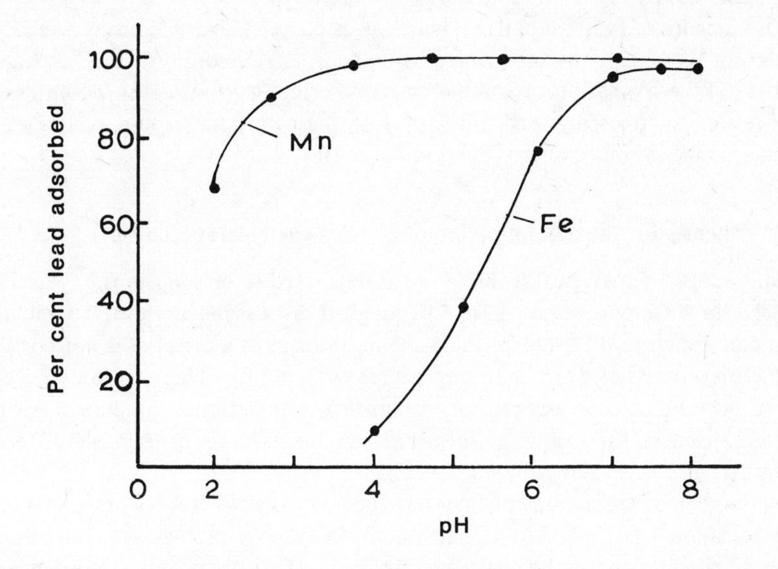

Fig. 3.8 Effect of pH on the adsorption of lead onto hydrous oxides of iron and manganese [29].
[Total Lead] = 10^{-5} $[MnO_2]_T = 4.4 \times 10^{-5}$ $[Fe(OH)_3]_T = 6.2 \times 10^{-5}$.
All concentrations in mol dm^{-3}.

soluble under anaerobic conditions. They may therefore release their associated metals when they encounter anaerobic waters and sediments. Nevertheless, it is likely that a concomitant production of hydrogen sulphide under these conditions will remove the lead from the soluble phase once again, this time as highly insoluble lead sulphide.

3.4.5 Mixing with sea water

Some of the consequences for lead speciation of mixing inland waters with sea water have already been noted. The results of Hem's study [27] (Section 3.4.2) suggest that lead would be released from ion exchange sites on sediments. Rohatgi and Chen [30] have indeed found a release of lead from sediments upon mixing with sea water (Table 3.14). However, their results showed that equilibrium was achieved only after 20 days, although a major part of the

Table 3.14 Percentage release of trace metals from suspended solids at equilibrium in mixtures with seawater, [30].

Solids	Lead released (%)
(a) Digested sludge	35
(b) Primary and mixture of primary and secondary effluents	53–58
(c) Dry weather flow from Los Angeles river	17

release occurred fairly rapidly. Their suggested release mechanisms were: (1) oxidation of organic matter or metal sulphides; (2) desorption from suspended solids; (3) the formation of soluble metal-chloride or metal-organic complexes. The results of this study probably apply to the behaviour of lead contained in sewage sludges dumped at sea (Section 3.2.1.2).

3.4.6 Schemes for the determination of physico-chemical speciation

A wide range of analytical techniques has been used to investigate the speciation of lead in different waters [31]. Although it is possible to separate and discriminate between different physico-chemical forms of a metal, it is not possible to identify specific forms with any degree of certainty. They can only be described in terms of their association with different operational components of the analysis. Hence, for example, filtration can be used to distinguish different groupings of species along the lines of Fig. 3.5.

A number of speciation schemes have been devised whereby a range of analytical techniques is applied to a single sample in order to provide as much information as possible about the different species. The first reasonably comprehensive scheme was developed by Batley and Florence [32]. The scheme has been applied mainly to the speciation of metals in sea water. The broad conclusion of their studies is that lead is associated largely with colloidal inorganic particles [33].

A somewhat different scheme has been developed for the speciation of lead in freshwater [34]. This scheme uses the size association of lead as the basic distinguishing feature (cf. Fig. 3.5). The lability of the lead (essentially the ease with which the lead complexes in the sample will dissociate) is also determined using both anodic stripping voltammetry (ASV) and Chelex-100 ion exchange resin. This provides an indication of the types of complexes present. Organic complexation is identified by irradiation with u.v. light which causes breakdown of the organic complexing agents.

The speciation of lead in two different types of tapwater, analysed according to this scheme, is summarized in Table 3.15. The results for iron are also included because of the potential importance of colloidal ferric hydroxide (Section 3.4.4). It is concluded on the basis of the whole range of measurements made on the samples [34] that in the Bentham tapwater, which is of low organic content, the lead is probably associated principally with colloidal ferric hydroxides. On the other hand, in the Glasgow tapwater, which is of higher organic content and with a lower iron concentration, the lead is mostly associated with organic complexes of high molecular weight (0.015 μm is equivalent to the molecular weight of *ca.* 300 000). These results serve to highlight the point that no one chemical form of lead will exist in all waters, but rather that lead speciation will be adapted to the chemical matrix of each particular water.

Table 3.15 Speciation of lead in two first flush tapwater samples (based on [34]).

Size fraction (μm)	Metal concentrations ($\mu g\ dm^{-3}$ %)			
	Bentham, N. Yorkshire*		Glasgow, Scotland†	
	Pb	Fe	Pb	Fe
>12	–	51(30)	3(1)	1.8(3)
1.0–12	16(19)	51(30)	49(13)	20.1(35)
0.4–1.0	14(17)	16(10)	7(2)	2.4(4)
0.08–0.4	22(26)	34(20)	59(16)	6.1(11)
0.015–0.08	8(10)	10(6)	198(53)	18.0(31)
<0.015	24(29)	6(4)	56(15)	9.0(16)
Total	84	168	372	57.4
ASV – labile‡	20(24)	NA§	236(63)	NA
Chelex – labile‡	39(46)	20(12)	207(56)	5.9(10)
Acid ASV – labile¶	87(104)	NA	382(103)	NA

*pH 6.8; alkalinity 20 mg dm^{-3} (as $CaCO_3$); organic carbon 1.6 mg dm^{-3}.
†pH 6.7; alkalinity 4 mg dm^{-3} (as $CaCO_3$); organic carbon 8.0 mg dm^{-3}.
‡1 μm filtrate.
§NA = not available.
¶Unfiltered sample 0.5% HNO_3.

References

[1] Carter, L. (1976), Marine Pollution and Sea Disposal of Wastes, *Chem. Ind.*, **19**, 825–9.

[2] Patterson, C., Settle, D., Schaule, B. and Burnett, M. (1976), Transport of Pollutant Lead to the Oceans and Within Ocean Ecosystems, in *Marine Pollutant Transfer* (ed. H. L. Windom, and R. A. Duce, Lexington Books, Lexington, pp. 23–38.

[3] Anderson, D. and Clark, R. (1979), Development of a Pretreatment Process for Toxic Metals Removal, *Proceedings of the Effluent Water Treatment Convention*, November 1978, Birmingham, UK, Brintex Exhibitions Ltd., London.

[4] Smith, B. C., Coppock, B. W., Scott, T. C. and Firkin, G. R. (1975), Pollution Control on an ISP Complex, *Proceedings of the South Australian Conference (Part A)*, Adelaide, Port Pirie, Australasian Inst. Mining Metal, Victoria, Australia, pp. 277–87.

[5] Prater, B. E. (1975), The Metal Content and Dispersion Characteristics of Steelworks' Effluents Discharging to the Tees Estuary, *Water Pollut. Control,* **74**, 63–78.

[6] Lester, J. N., Harrison, R. M. and Perry, R. (1979), The Balance of Heavy Metals Through a Sewage Treatment Works. I. Lead, Cadmium and Copper, *Sci. of the Tot. Environ.*, **12**, 13–23.

[7] Helsel, D. R., Kim, J. I., Gizzard, T. J., Randall, C. W. and Hoehn, R. C. (1977), Land Use Influences on Metal Yields in Storm Drainage, paper presented at *50th Annual Conference of the Water Pollution Control Federation* October 1977, Philadelphia, Pennsylvania.

[8] Thorell, L. (1977), Pollutants from Swedish Municipal and Industrial Outlets into the Baltic Sea, *Ambio Spec. Rep.*, **5**, 219–24.

[9] Norton, M. G. (1976), The Operation of the Dumping at Sea Act 1974, *Chem. Ind.*, 19, 829–34.

[10] Jones, R. E. (1978), Heavy Metals in the Estuarine Environment, *Water Res. Cent. Tech. Rep.*, **TR 73**, Water Research Centre, Medmenham, Bucks, pp. 128.

[11] Shaheen, D. G. (1975), Contributions of Urban Roadway Usage to Water Pollution, *Environ. Prot. Agency Rep., EPA 600/2-75-004*, Washington, pp. 228.

[12] Laxen, D. P. H. and Harrison, R. M. (1977), The Highway as a Source of Water Pollution: An Appraisal with the Heavy Metal Lead, *Water Res.*, **1**, 1–11.

[13] Water Research Centre (1977), Pollution from Urban Run-off, *Notes on Water Res.*, **12**, Stevenage, Herts, pp. 4.

[14] Hedley, G. and Lockley, J. C. (1975), Quality of Water Discharged from an Urban Motorway, *Water Pollut. Control*, **74**, 659–74.

[15] Bryan, E. H. (1974), Concentrations of Lead in Urban Stormwater, *J. Water Pollut. Control Fed.*, **46**, 2419–21.

[16] Andren, A. W., Lindberg, S. E. and Bate, L. C. (1975), Atmospheric Input and Geochemical Cycling of Selected Trace Elements in Walker Branch Watershed, *Oak Ridge Natl. Lab. Environ. Sci. Div. Publ.*, **728**, 68.

[17] Patterson, J. W. and Minear, R. A. (1973), *Wastewater Treatment Technology*, 2nd Edn, US NTIS PB Rep., PB 216, 162, Nat. Tech. Info. Service, Springfield, Va.

[18] Rimer, A. E. and Reynolds, D. E. (1978), Characterization and Impact of Stormwater Runoff from Various Land Cover Types, *J. Water Pollut. Control Fed.*, **50**, 252–64.

[19] Harrison, R. M., Perry, R. and Wellings, R. A. (1975), Lead and Cadmium in Precipitation: Their Contribution to Pollution, *J. Air Pollut. Control Assoc.*, **25**, 627–30.

[20] Aston, S. R. and Thornton, I. (1977), Regional Geochemical Data in Relation to Seasonal Variations in Water Quality, *Sci. Tot. Environ.*, **7**, 247–60.

[21] Abdullah, M. I. and Royle, L. G. (1973), The Occurrence of Lead in Natural Waters, *International Symposium on the Environmental Health Aspects Lead*, Amsterdam 1972, Commission of the European Community, Luxemburg, pp. 113–24.

[22] Patterson, C. (1977), Personal communication.

[23] Lovering, T. G. (Ed.) (1976) Lead in the Environment, *US Geol. Surv. Prof. Paper*, US Government Printing Office, Washington, 957, 90.

[24] Webb, J. S. (1978), *The Wolfson Geochemical Atlas of England and Wales*, Clarendon Press, Oxford.

[25] Ainsworth, R. G., Bailey, R. J., Commins, B. T., Packham, R. F. and Wilson, A. L. (1977), Lead in Drinking Water, *Water Res. Cent. Tech. Rep.*, **TR 43**, Water Research Centre, Medmenham, Bucks, p. 34.

[26] Stumm, W. and Bilinski, H. (1973), Trace Metals in Natural Waters:Difficulties of Interpretation Arising from our Ignorance of their Speciation, in *Advances in Water Pollution Research, Proceedings of the 6th International Conference*, Jerusalem, 1972, Pergamon Press, Oxford, pp. 39–52.

[27] Hem, J. D. (1976), Geochemical Controls on Lead Concentration in Stream Water and Sediments, *Geochim. Cosmochim. Acta.*, **40**, 599–609.

[28] Lee, G. F. (1975), Role of Hydrous Metal Oxides in the Transport of Heavy Metals in the Environment, in *Heavy Metals in the Aquatic Environment* (ed. P. A. Krenkel), Pergamon Press, Oxford, pp. 137–147.

[29] Gadde, R. R. and Laitinen, H. A. (1974), Studies of Heavy Metal Adsorption by Hydrous Iron and Manganese Oxides, *Anal. Chem.*, **46**, 2022–26.

[30] Rohatgi, N. and Chen, K. Y. (1975), Transport of Trace Metals by Suspended Particulates on Mixing with Sea water, *J. Water Pollut. Control Fed.*, **47**, 2298–2316.

[31] Harrison, R. M. and Laxen, D. P. H. (1980), Metals in the Environment, I – Chemistry, *Chem. Brit.*, **16**, 316–20.

[32] Batley, G. E. and Florence, T. M. (1976), A Novel Scheme for the Classification of Heavy Metal Species in Natural Waters, *Anal. Lett.*, **9**, 379–88.

[33] Florence, T. M. and Batley, G. E. (1980), Chemical Speciation in Natural Waters, *CRC Crit. Rev. Anal. Chem.* in press.

[34] Harrison, R. M. and Laxen, D. P. H. (1980), Physico-chemical Speciation of Lead in Drinking Water, *Nature*, **286**, 791–93.

4

Lead in soils

4.1 Introduction

When incorporated in the soil, lead is of very low mobility. Hence once contaminated, a soil is liable to remain polluted with lead. This might have adverse consequences for soil fertility if the degree of contamination is great. In addition, plants grown on lead-rich soils incorporate lead and thus the concentration of lead in crop plants may be increased slightly. These questions are explored in the following sections.

4.2 Sources of lead in soil

4.2.1 Parent materials

The parent geological materials from which soils are derived are important contributors of lead. In immature soils the lead content often relates well to that in the parent material, but in more developed soils this relationship may be lost as pedogenetic factors influence the distribution of lead in the soil profile [1]. Thus the mean level of lead in unpolluted soils approximates to the concentration in the Earth's upper lithosphere.

As a consequence of lead mineralization, substantially elevated levels of lead may be found in some local soils.

4.2.2 Dry deposition of airborne lead

Airborne lead is subject to dry deposition by two major mechanisms. The first is that of gravitational settling, and this is significant for all particles of $>10\ \mu m$, but is rapid only for the very large particles ($>50\ \mu m$) [2]. The second mechanism, which applies to particles of all sizes but is of greatest importance for the smaller particles, is deposition onto surfaces due to impaction, and for the extremely small particles due also to Brownian diffusion.

The magnitude of particle deposition due to gravitational settling from the plume of an elevated point source, such as smelter stack, is a function of the atmospheric concentration and the settling velocity, itself determined by Stoke's Law. A description of the means of estimating deposition by this mechanism as a function of the distance from the point of emission is given by Harrison and Parker [2] who also describe the techniques of sampling dry deposition.

Deposition of vehicle exhaust lead aerosol to plant and soil surfaces by diffusion and impaction mechanisms has been studied by Little and Wiffen [3]. These workers found that rough or hairy leaf surfaces were up to eight times more efficient at collection of lead than were smooth surfaces. The velocity of deposition, V_g, of lead aerosol was determined and some results are shown in Fig. 4.1.

$$V_g = \frac{\text{deposition (g cm}^{-2}\text{ s}^{-1})}{\text{mean atmospheric concentration (g cm}^{-3})}.$$

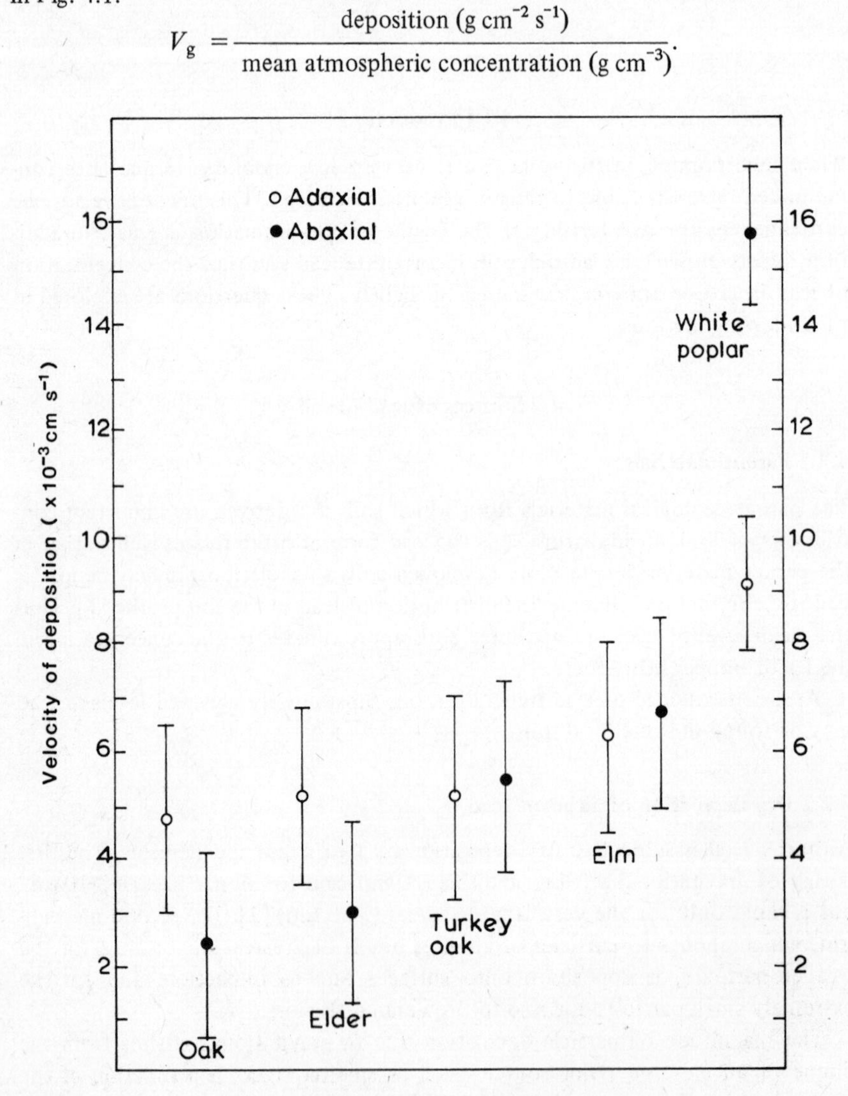

Fig. 4.1 Velocity of deposition of ^{203}Pb-labelled exhaust to adaxial (upper) and abaxial (lower) leaf surfaces (mean values ± 95% confidence limits) [3].

Fresh exhaust aerosols showed higher values of V_g than aged aggregated aerosols, and this was attributed to the higher diffusivity of the finer primary exhaust particles [3].

Deposition velocities to grass and soil surfaces were also measured (Table 4.1). Grass is clearly a rather efficient aerosol collector; more efficient than soil.

Table 4.1 Mean deposition velocities of ^{203,212}Pb-labelled exhaust to grass and soil surfaces [3].

	Aggregated Aerosol	Non-aggregated (primary) Aerosol
Grass (cm s^{-1})	0.0188	0.1273
Soil under grass (cm s^{-1})	0.0067	0.0146
Bare soil (cm s^{-1})	0.0081	0.0354
Grass and soil (cm s^{-1})	0.0255	0.1419
Percentage of total catch intercepted by grass	73.7%	89.7%
Ratio (grass + soil)/bare soil	3.15	4.01
Ratio bare soil/soil under grass	1.21	2.42

Lead deposited onto vegetation reaches the soil sooner or later. Much is rapidly washed onto the soil by rainwater. Little and Wiffen [3] found that 50% of deposited lead could be washed from a leaf surface with one rinse of distilled water. Similarly rainfall removed 30–50% of deposited lead. Even lead which is not rapidly transferred to the soil in this manner will eventually find its way there upon the death and decomposition of the plant.

The magnitude of deposition of vehicle-emitted lead is not readily estimated. Shaheen [4] measured seasonal average rates of deposition on the road itself of 4.1–11.0 $\times$ 10^{-3} g (Pb) axle-km^{-1} at a number of sites in Washington DC. These figures are in good accord with the report of Ter Haar *et al.* [5] that over a full Federal test cycle, emission of coarse lead particles amounted to 2.8 $\times$ 10^{-2} g mile^{-1} (8.8 $\times$ 10^{-3} g axle-km^{-1}). These figures are likely to depend heavily upon the driving mode, as this influences not only the lead emission as a percentage of that burned, but also the particle size distribution. Sampling next to a motorway in Britain at a site where cars were cruising steadily, Little and Wiffen [3] found that only 22% of emitted lead was deposited within 100 m of the carriageway, the remainder staying airborne for longer distances.

4.2.3 Wet deposition of airborne lead

Scavenging by precipitation is an efficient means of removing particulate pollutants from the atmosphere. There are two main processes involved: rainout, which is the scavenging by water droplets within the cloud layer, and washout, the scrubbing of air by falling raindrops (or snow). Precipitation scavenging is most effective for pollutants having a substantial vertical dispersal in the atmosphere, and the influence of local sources upon wet deposition is unlikely to be as great as in the instance of dry deposition which is greatly enhanced by the presence of local emissions.

It is only recently that samplers have been developed which discriminate dry and wet deposition. From these, it is known that at background sites, remote from major sources of emission, dry and wet deposition each account for about a half of the total deposition of lead.

Lead deposition at background sites in Britain has been measured as 0.046–0.186 mg m^{-2} day^{-1} [6]. This may be compared with the values determined at sites in and around London [7], shown in Table 4.2. Also included in this table is a value of lead deposition determined in Newcastle-upon-Tyne, England, in 1933, presumably as a result primarily of the burning of coal. This highly elevated level may be contrasted with the natural rate of 0.0044 mg m^{-2} day^{-1}, estimated from the analysis of lake sediments [8].

Table 4.2 Total deposition of lead at various sites.

Site	Deposition surface	Pb deposited (mg m^{-2} day^{-1})	Reference
London, SW7	Deposit gauge	0.230	[7]
Hyde Park, London	Deposit gauge	0.119	[7]
London, SE2	Deposit gauge	0.188	[7]
Hounslow, London	Deposit gauge	0.128	[7]
Heathrow Airport (airside)	Deposit gauge	0.086	[7]
Heathrow Airport (landside)	Deposit gauge	0.135	[7]
Cambridgeshire (rural site)	Deposit gauge	0.028	[7]
Central reserve of M4 motorway	Grass and soil	10.2–15.9	[3]
0 m from hardshoulder of M4	Grass and soil	5.8–15	[3]
5 m from hardshoulder of M4	Grass and soil	2.3–6.2	[3]
15 m from hardshoulder of M4	Grass and soil	1.1–3.1	[3]
30 m from hardshoulder of M4	Grass and soil	0.6–1.8	[3]
Avonmouth smelter (0.1 km)	Deposit gauge	5.0–11.5	cited in [3]
Avonmouth smelter (1 km)	Deposit gauge	0.5–4	cited in [3]
Avonmouth smelter (3 km)	Deposit gauge	0.4–1	cited in [3]
Newcastle-upon-Tyne, 1933	Deposit gauge	2.49	cited in [7]
UK background	Deposit gauge	0.046–0.186	[6]
Estimated natural		0.0044	cited in [8]

4.2.4 Disposal of sewage sludge to land

Sewage sludges are the solids separated during the treatment of waste waters. As produced by the sewage treatment works they are comprised of more than 90% water, and are sometimes dried or filtered prior to disposal. In the sewage treatment process, 80–100% of the input lead is typically incorporated in the sludges [9], along with other heavy metals present in the raw sewage. These sludges are rich in nitrogen and phosphorus, and so as to make use of these important plant nutrients and to ease the problems of sludge disposal, considerable volumes of sewage sludge are currently spread on the land.

Typical concentrations of lead in sewage sludges are shown in Table 4.3. In most countries guidelines exist to control the disposal of sewage sludges to land. These are normally based upon the zinc, copper and nickel content of the sludge and not the lead content. Hence considerable quantities of lead may be added to the land over a normal 30 year disposal period. Estimates based upon a typical range of sewage sludge compositions are included in Table 4.3.

Table 4.3

Lead in sewage sludges						Reference
Total content (mg kg^{-1} dry material)						
Range	mean	median				
120–3000	820	700				[10]
Number of samples within content ranges (mg kg^{-1} dry weight)						
0–1	1–10	10–100	100–1000	1000–10 000	> 10 000	[10]
0	0	0	36	6	0	
Estimated addition to land in sewage sludge (kg ha^{-1} y^{-1})						
Digested sewage sludge (domestic)*			Digested sewage sludge (industrial)†			
0.59			0.07–0.99			[7]

*Calculated from figures provided by the Greater London Council for the metal content of digested sewage sludge from domestic sources, on the basis of annual additions of sludge over a 30 year period (maximum permissible addition).

†Digested sewage sludges containing industrial wastes. A 30 year disposal period assumed.

A recent revision of the rules governing the disposal of sewage sludge to the land in the UK has placed some restriction on the quantity of lead added to the land. The guideline is based upon the premises that lead is not readily taken up and translocated by plants, and hence elevated levels of lead in soil will have little effect upon the content of the plants (normally about 3 mg kg^{-1} dry weight), and secondly that lead does not produce toxic effects upon plant growth.

The restriction upon disposal of lead-rich sewage sludges arises from the likelihood of direct ingestion by animals of grass or soil contaminated by sewage sludge. It is consequently recommended that where sludges might accumulate upon the grass or soil surface and hence be consumed by animals, only sludges of lead content <2000 mg kg^{-1} dry weight should be disposed. Otherwise, where the sludge is to be mixed with the soil, the total lead disposed over the normal 30 year period should be limited to 1000 kg ha^{-1}, corresponding to about 450 mg kg^{-1} of soil of 200 mm depth.

4.2.5 Other sources of lead contamination of soils

Accidental irrigation or flooding with water of a high lead concentration may occasionally occur. Such a water might derive from worked, mineralized zones or the leaching of spoil heaps [11]. The tipping of lead-containing solid wastes such as fly ash may occur [11]. In this particular instance however, so little of the lead is water-soluble that this does not create a problem [11].

4.3 Concentrations of lead in soil

As indicated above, mean concentrations of lead in unpolluted soils relate closely to concentrations in the parent geological material. Consequently in non-mineralized areas (most areas, in practice), concentrations range from 2 to 200 mg kg^{-1}, with most samples being in the range 5 to 25 mg kg^{-1} [1]. Concentrations in excess of these are generally indicative of lead pollution or mineralization. In areas previously worked for mineral deposits, concentrations may reach 20 000 mg kg^{-1}, but are more typically within the range of 1000 to 2000 mg kg^{-1} [11].

Since soils are strong accumulators of lead, the analysis of lead in soil is an excellent indicator of accumulated deposition in the vicinity of a source of the metal. In one survey around a secondary smelter [12], concentrations of lead up to 21 000 mg kg^{-1} (dry weight) were found in the upper 5 cm of soil adjacent to the smelter with the levels decreasing exponentially with distance from the source. Mean concentrations of lead in soil around the Silver Valley lead smelter [13] (air lead concentrations in Fig. 2.3) are shown in Fig. 4.2. Changes in soil lead between the 1974 and 1975 surveys presumably arise mainly from random errors introduced by minor spatial variability in the lead concentration.

There have been numerous reports on the concentrations of lead in roadside soils. Typical distributions with respect to the road are shown in Fig. 4.3, which shows analysis of soil collected alongside a road of low traffic volume in New Zealand [14]. Two features are worthy of note. Firstly, deposited lead is normally restricted to the first few centimetres of an undisturbed soil, and secondly concentrations fall to almost background values within 100 m of the road. The diminution of lead concentration with distance from the road is approximately exponential, and is shown in Fig. 4.4 for roads of differing traffic volume.

In comparing data on lead concentrations in soils, it is important to take account of the techniques of extraction and analysis used. Some techniques are designed to extract only 'available' lead, i.e. that supposedly available for plant uptake, and even those designed to determine the total lead are of substantially differing efficiencies [16].

4.4 Uptake of soil lead by living organisms

Concentrations of lead in roadside grass may reach several hundred mg kg^{-1} (dry weight) but the bulk of this lead arises from surface deposition, and not from uptake from the soil. The degree of uptake from the soil by growing plants

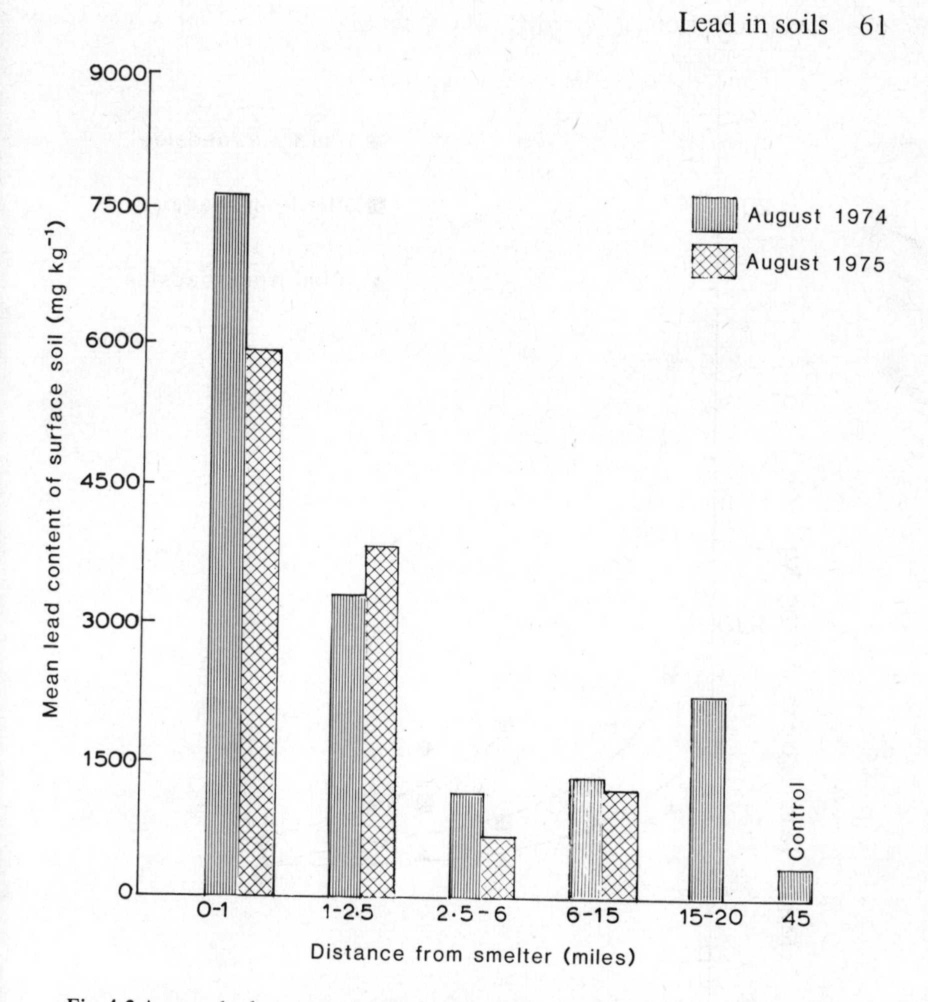

Fig. 4.2 Average lead content of surface soil, by area, in the Silver Valley study [13].

is highly variable, being dependent upon the availability of the lead. Factors such as a low cation exchange capacity, low organic content and low pH cause enhanced lead mobility and hence increased availability for uptake.

Many plants absorb considerable amounts of lead into the roots, but translocate only limited quantities into the above-ground portions. Table 4.4 shows the relationship between the lead content of lettuce and radish grown in contaminated soils and the concentration in the soil, in the absence of atmospheric contamination. In this instance lead concentrations in the soil are expressed as both total lead and available lead (AL-soluble Pb), the latter being defined as that soluble in an ammonium acetate–lactate solution [17]. In this study a significant correlation ($P < 0.05$) between lead concentration in plants and AL-soluble lead in soil was found.

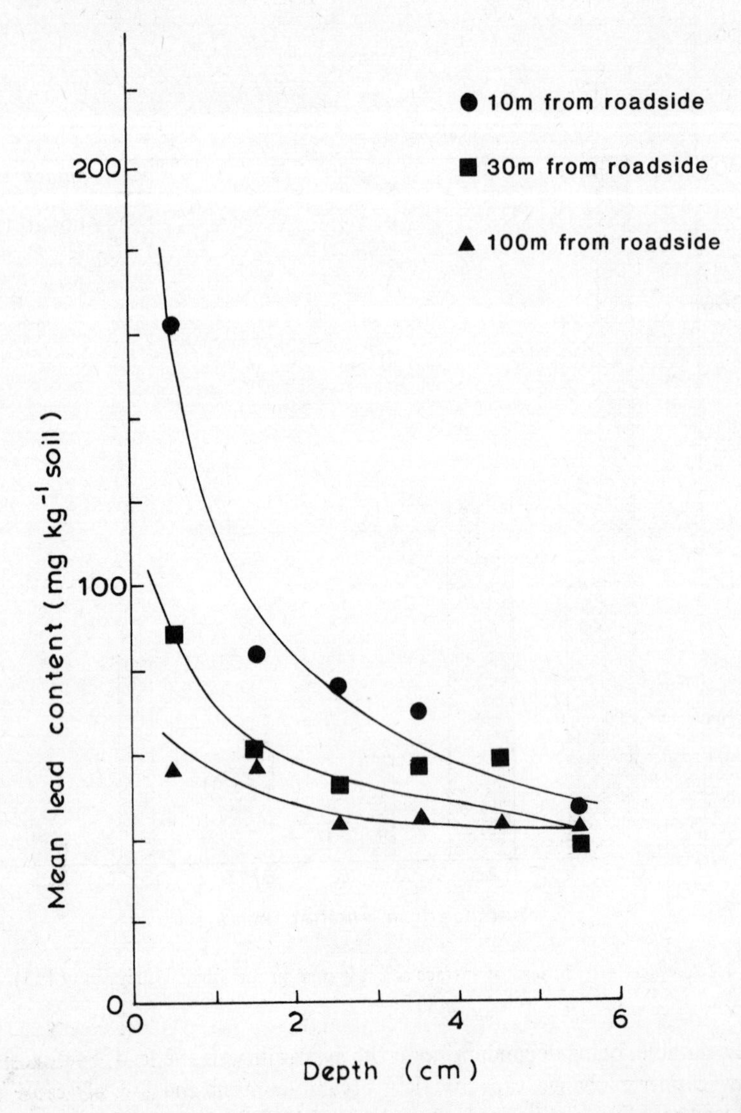

Fig. 4.3 Lead content of soils, as a function of depth, at various distances from a highway with low traffic volume [14].

The lead deposited in soils and vegetation can cause enhanced levels of lead in soil invertebrates. Earthworms have been studied extensively and have been found to accumulate several hundred mg kg^{-1} of lead (dry weight) in extreme cases. Concentrations of lead in earthworms show correlation with the concentrations in roadside soil at the point of sampling. Small mammals living in the

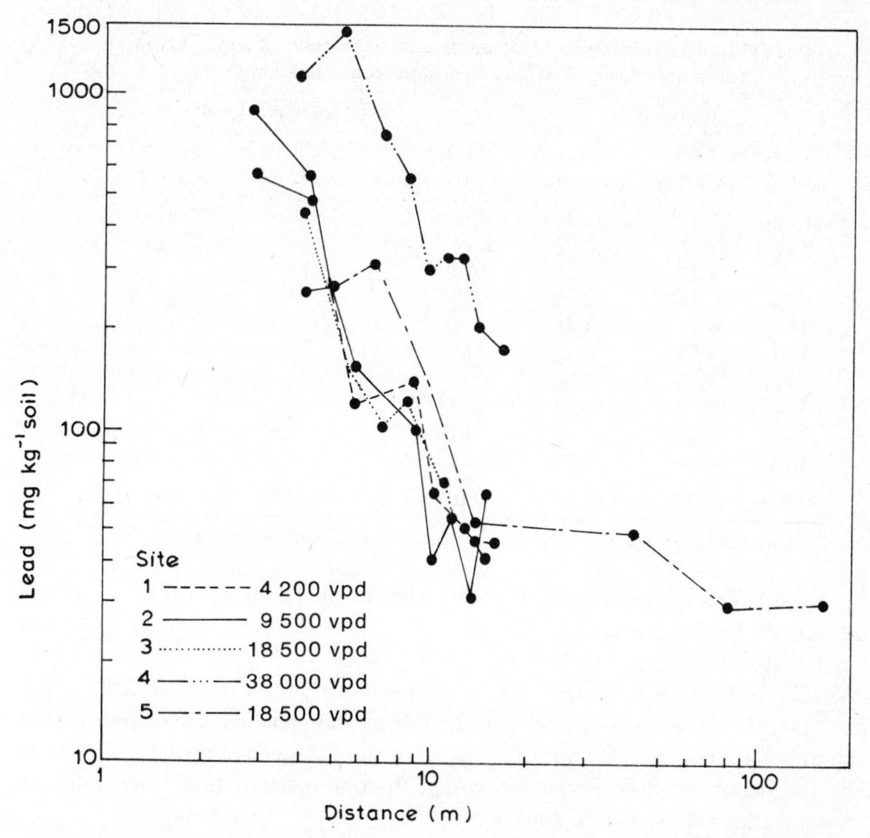

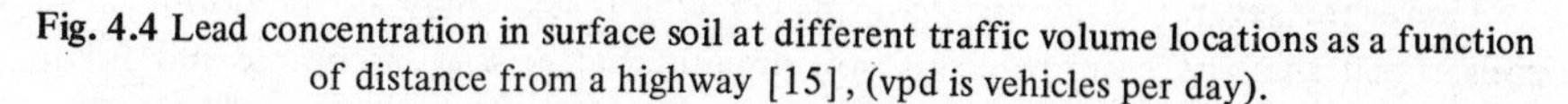

Fig. 4.4 Lead concentration in surface soil at different traffic volume locations as a function of distance from a highway [15], (vpd is vehicles per day).

roadside environment, such as mice and voles, also show elevated levels of lead in specific organs, related to traffic volume and distance from the highway.

One point of concern has been the possible absorption of lead ingested by ruminant animals (cattle and sheep), with subsequent incorporation into the muscle or milk. It appears, however, that such animals absorb lead in the alimentary tract only very inefficiently, about 99% being excreted in the faeces [11]. Hence concentrations of lead in meat and milk can be very little affected. The mean concentrations of lead in meat and fish in the UK is 0.17 mg kg^{-1} (range <0.01–0.70 mg kg^{-1}) and in milk 0.03 mg kg^{-1} (range <0.01–0.08 mg kg^{-1}) [11].

4.5 Chemistry of lead in soils

This is an extremely complex subject, not yet fully understood. Any remarks here can serve only as a crude introduction.

Table 4.4 Lead concentration in the upper parts of lettuce and radishes grown in contaminated soils remote from aerial contamination [17].

	Radishes		Contaminated soils		
Lettuce $\mu g (Pb) g^{-1}$	Leaves $\mu g (Pb) g^{-1}$	Stems $\mu g (Pb) g^{-1}$	Total Pb mg (Pb) kg^{-1}	AL-soluble Pb mg (Pb) kg^{-1}	AL-Pb:Total Pb (%)
11.02	10.47	11.04	3055	556	18.2
8.82	11.72	15.00	4644	624	13.4
–	5.27	–	910	55	6.0
5.98	3.58	2.78	799	63	7.9
3.92	4.77	1.46	672	35	5.2
3.24	4.53	–	685	23	3.4
9.63	9.96	3.80	944	143	15.1
3.10	5.62	3.44	873	33	3.7
2.85	4.43	–	361	8.6	2.3
4.18	6.88	1.36	2289	134	5.8
(2.63)	(3.37)	(1.56)	(40)	(7.0)	(5.75)*

*Figures in parentheses are representative of urban background values.

Lead added to soil may react with available soil anions such as SO_4^{2-}, PO_4^{3-} or CO_3^{2-} to form sparingly soluble salts [8]. Compounds such as basic lead carbonate, $Pb_3(OH)_2(CO_3)_2$ (log K_s = −46.8 at 25° C) and chloropyromorphite, $Pb_5(PO_4)_3Cl$, (log K_s = −84.4) provide the least soluble simple inorganic salts at near-neutral pH. In fact, X-ray powder diffraction (XRD) studies by Olsen and Skogerboe [18] have shown the more soluble $PbSO_4$ to be the major crystalline lead compound in contaminated soils. The results of these workers are shown in Table 4.5.

There are several other mechanisms by which lead may be immobilized in soils. Lead may be complexed by soil organic matter such as humic and fulvic acids which are themselves adsorbed onto soil solids. Adsorption of lead onto clay mineral particles and onto hydrous oxides of iron and manganese has been shown to be effective. Manganese dioxide appears to be the strongest adsorber of lead, adsorbing this metal in preference to other heavy metals [8]. Soils have a cation exchange capacity, and hence lead may be immobilized by ion exchange at sites on the solid material. This may facilitate attachment of substantial amounts of lead, but cannot account entirely for the immobility of lead in soils. A recent report [19] concludes that the majority of lead immobilized in soil is associated with organic matter.

The actual mechanism of lead association with soils is no doubt a mixture of some or all of the above processes, the predominant mode probably depending upon the composition and pH of the specific soil in question.

4.6 Lead in street dusts

Street dust is the solid material which accumulates in the gutters at the edge of paved roads. Many reports on the lead content of such dusts have been published. Work in Manchester showed concentrations of a similar order throughout the city, irrespective of the sampling site. In contrast, samples from Lancaster, a small town in the north of England, showed a distinct relationship between the sampling site and the lead concentration. These data, shown in Table 4.6 are typical of lead concentrations in street dusts in areas without proximate industrial lead emissions [20]. Levels of organic lead (dialkyl, trialkyl plus tetraalkyllead compounds) in the dusts were also measured and found to be small; in 46 samples, concentrations ranged from 0.4 to 7.4 mg(Pb) kg^{-1} of the total weight of dust, representing $<1\%$ of total lead [21].

The results of X-ray powder diffraction analyses of street dusts collected in Lancaster, Birmingham and London are summarized in Table 4.7. Elemental lead (Pb^0) was found only in samples collected in car parks and may arise from chemical reduction of PbO by CO in a cold, choked engine. In common with Olson and Skogerboe's results for soils (Table 4.5), lead sulphate was the most frequently encountered as a crystalline compound of lead. This is formed by rainwater leaching of $PbSO_4 \cdot (NH_4)_2SO_4$ deposited from the atmosphere [22] (Section 2.5.2). Extensive weathering may be the cause of formation of $2PbCO_3 \cdot Pb(OH)_2$, the compound predicted as the least soluble crystalline form of lead in natural waters (Chapter 3). In their speciation studies of street dusts, Biggins and Harrison [22] noted that only a small proportion of the lead was present in crystalline forms susceptible to XRD analysis. Since street dusts appear to be composed largely of soil, it is probable that crystalline lead salts are progressively dissolved by rainwater and redeposited in the non-crystalline chemical forms typical of soils (Section 4.5).

Highly elevated concentrations of lead may be found in deposited dust and in street dusts close to lead works. Concentrations of up to 300 000 mg kg^{-1} (30%) lead have been reported, but typical values are of the order of 1000–50 000 mg kg^{-1} [11].

When viewing these concentrations, one point worthy of note is that lead levels in deposited urban dusts sampled in 1928, before the widespread use of leaded petrol, showed concentrations very similar to those found today in street dusts [11]. Presumably the far greater emissions of smoke from coal combustion and the use of high-lead paints had a substantial influence upon lead deposition.

Concentrations of lead in household dusts are of a similar order to those in street dusts, although it is not clear whether the prime source of lead is within or without the house [20].

Table 4.5 Crystalline lead compounds in contaminated soils [18].

Soil lead preconcentration results

Sample identity	Lead concentration, ($\mu g\ g^{-1}$) (percent of total lead present)			
	Before separation	Dense fraction*	Magnetic fraction*	Nonmagnetic fraction*
Fort Collins-1	2000	66 500 (75.4)	47 100 (25.4)	81 600 (50.0)
Fort Collins-2	1900	58 600 (60.3)	47 700 (19.9)	66 200 (40.4)
Denver-1	960	24 200 (79.2)	14 500 (26.8)	45 200 (52.4)
Chicago-10	7000	120 800 (65.6)	106 400 (47.1)	184 700 (18.5)
Chicago-20	2100	27 800 (33.4)	11 300 (12.8)	127 100 (20.6)
Chicago-30	4800	75 900 (74.6)	42 400 (30.8)	168 700 (43.8)
Chicago-40	2400	46 700 (70.0)	22 700 (39.9)	83 500 (30.1)
Missouri-1	1540	44 300 (15.8)	21 000 (1.4)	47 700 (14.4)

Lead compounds identified by X-ray powder diffraction

Sample identity	Soil fraction	Compounds found	Concentration estimates†
Fort Collins-1 and Fort Collins-2	Magnetic	$PbSO_4$	Major
	Nonmagnetic	$PbSO_4$	Major
		$PbO \cdot PbSO_4$	Minor
		PbO_2	Trace
		PbO‡	Trace

Denver-1	Magnetic	$PbSO_4$	Major
	Nonmagnetic	$PbSO_4$	Major
Chicago-10	Magnetic	$PbSO_4$	Major
	Nonmagnetic	Pb^0	Major
		$PbSO_4$	Minor
Chicago-20, 30 and 40	Magnetic	$PbSO_4$	Major
	Nonmagnetic	$PbSO_4$	Major
Missouri-1	Magnetic	None §	–
	Nonmagnetic	PbS	Major
		$PbSO_4$	Minor

* Density greater than 3.32 g cm^{-3}.

† Major indicates the principal portion of lead present in the soil fraction indicated and therefore the principal portion of the soil sample; minor refers to approximately 1–10% of the Pb in the respective fractions; trace quantities are less than approximately 1% of the total in each fraction.

‡ Assignment based on the presence of only the most intense *d*-spacing and therefore questionable.

§ Complex *d*-spacing pattern obtained with all intensities low; positive assignment of any one compound or group of compounds questionable.

Table 4.6 Concentrations of total lead in street dusts in Lancaster, England (mg kg^{-1} as collected) [20].

Site	Number of samples	Range of concentrations	Mean	Standard deviation
Car parks	4	39 700–51 900	46 300	5 900
	16	950–15 000	4 560	3 700
Garage forecourts	2	44 100–48 900	46 500	–
	7	1 370–4 480	2 310	1 150
Town centre streets	13	840–4 530	2 130	960
Main roads	19	740–4 880	1 890	1 030
Residential areas	7	620–1 240	850	230
Rural roads	4	410–870	570	210

Table 4.7 Crystalline compounds of lead identified in street dusts* [22].

Lead compound	Number of samples
$PbSO_4$	4
Pb^0	3 (car parks)
$PbSO_4 \cdot (NH_4)_2SO_4$	2
Pb_3O_4	1
$PbO \cdot PbSO_4$	1
$2PbCO_3 \cdot Pb(OH)_2$	1

*A total of 19 samples from Lancaster, Birmingham and London were analysed.

References

[1] US Geological Survey (1976), *Lead in the Environment*, Professional paper No. 957, Washington DC.

[2] Harrison, R. M. and Parker, J. (1977), Analysis of Particulate Pollutants, in *Handbook of Air Pollution Analysis* (ed. R. Perry and R. J. Young),Chapman and Hall, London, pp. 84–156.

[3] Little, P. and Wiffen, R. D. (1977), Emission and Deposition of Petrol Engine Exhaust Pb – I. Deposition of Exhaust Pb to Plant and Soil Surfaces, *Atmos. Environ.*, **11**, 437–47.

[4] Shaheen, D.G. (1975), Contributions of Urban Roadway Usage to Water Pollution, *Environ. Prot. Agency Rep.*, EPA 600/2–75–004, Washington DC.

[5] Ter Haar, G. L., Lenane, D. L., Hu, J. N. and Brandt, M. (1972), Composition, Size and Control of Automotive Exhaust Particulates, *J. Air Pollut. Control Assoc.*, **22**, 39–46.

[6] Cawse, P. A. (1974), A Survey of Atmospheric Trace Elements in the UK (1972–73), *UK Atomic Energy Authority Rep.*, AERE R7669, HMSO, London.

[7] Harrison, R. M., Perry, R. and Wellings, R. A. (1975), Lead and Cadmium in Precipitation: Their Contribution to Pollution, *J. Air Pollut. Control Assoc.*, **25**, 627–30.

[8] Laxen, D. P. H. and Harrison, R. M. (1977), The Highway as a Source of Water Pollution: An Appraisal with the Heavy Metal Lead, *Water Res.*, **11**, 1–11.

[9] Lester, J. N., Harrison, R. M. and Perry, R. (1979), The Balance of Heavy Metals through a Sewage Treatment Works. I. Lead, Cadmium and Copper, *Sci. Tot. Environ.*, **12**, 13–23.

[10] Berrow M. L. and Webber, J. (1972), Trace Elements in Sewage Sludges, *J. Sci. Food Agric.*, **23**, 93–100.

[11] Central Unit on Environmental Pollution, Department of the Environment (1974), *Lead in the Environment and its Significance to Man*, HMSO, London.

[12] Linzon, S. N., Chai, B. L., Temple, P. J., Pearson, R. G. and Smith, M. L. (1976), Lead Contamination of Urban Soils and Vegetation by Emissions from Secondary Lead Industries, *J. Air Pollut. Control Assoc.*, **26**, 650–4.

[13] Yankel, A. J., von Lindern, I. H. and Walter, S. D. (1977), The Silver Valley Lead Study: The Relationship between Blood Lead Levels and Environmental Exposure, *J. Air Pollut. Control Assoc.*, **27**, 763–7.

[14] Ward, N. I., Reeves, R. D. and Brooks, R. R. (1975), Lead in Soil and Vegetation along a New Zealand State Highway with Low Traffic Volume, *Environ. Pollut.*, **9**, 243–51.

[15] Welch, W. R. and Dick, D. L. (1975), Lead Concentrations in Tissues of Roadside Mice, *Environ. Pollut.*, **8**, 15–21.

[16] Harrison, R. M. and Laxen, D. P. H. (1977), A Comparative Study of Methods for the Analysis of Total Lead in Soils, *Water Air Soil Pollut.*, **8**, 387–92.

[17] Kerin, Z. (1975), Relationship between Lead Content in the Soil and in the Plants contaminated by Industrial Emissions of Lead Aerosols, *Proceedings of the International Conference on Heavy Metals in the Environment*, Vol. II (2), Toronto, pp. 487–502.

[18] Olsen, K. W. and Skogerboe, R. K. (1975), Identification of Soil Lead Compounds from Automotive Sources, *Environ. Sci. Technol.*, **9**, 227–30.

[19] Zimdahl, R. L. and Skogerboe, R. K. (1977), Behaviour of Lead in Soil, *Environ. Sci. Technol.*, **11**, 1202–7.

[20] Harrison, R. M. (1979), Toxic Metals in Street and Household Dusts, *Sci. Tot. Environ.*, **11**, 89–97.

[21] Harrison, R. M. (1976), Organic Lead in Street Dusts, *J. Environ. Sci. Health*, **A11**, 417–23.

[22] Biggins, P. D. E. and Harrison, R. M. (1980), Chemical Speciation of Lead Compounds in Street Dust, *Environ. Sci. Technol.*, **14**, 336–9.

5

Control of lead in air

5.1 Industrial sources of lead

It has been indicated in earlier chapters that industrial emissions of lead contribute little on a nationwide basis to lead in the air, in comparison with automotive sources. This is largely a result of the very tight control exercised over industrial sources of lead, using techniques which will be described in this chapter. Nevertheless, as shown in Chapter 2, industrially emitted lead may have a considerable local effect upon air quality, and if ambient air quality standards such as that in the USA (1.5 $\mu g\ m^{-3}$) are to be met, then considerable improvements in emission control installations will be required.

Before methods of pollution control can be discussed, it is pertinent to examine the processes from which lead emissions arise. These will be covered in the following sections. It is especially important to consider the nature of the effluent gas streams (e.g. temperature, lead particle size distribution) in order that an appropriate control method may be selected.

5.1.1 Mining, smelting and refining of non-ferrous metals

Lead containing ores usually also contain zinc, copper and cadmium in varying proportions, the metals normally being present as sulphides, sulphates or oxides. The first processes in metal production are mining, crushing and grinding of the ores. US emissions from these processes and from smelting are shown in Table 5.1 and emission factors for the different processes are tabulated in Chapter 2. Emissions of lead are primarily fugitive (meaning lost in transfer or storage or due to leakage, rather than in defined, controlled streams) and control is by enclosure of a stock or process, or wetting to limit wind-blown losses. Particle sizes at this stage are rather large, and this fact combined with the high density of the ores means that emitted material has a rather small range of travel (Chapter 2).

After concentration by settling and flotation, the ore is transferred to the smelter.

Table 5.1 Lead emissions in the US in 1975 from primary non-ferrous metals production [1].

Process	Emission (tonnes)
Ore mining, crushing and grinding	493
Primary lead smelting	400
Primary zinc smelting	112
Primary copper smelting	1314
Total	2319

5.1.1.1 Primary smelting of lead

A flow diagram of a typical lead smelter is shown in Fig. 5.1. The first process of sintering the ore involves roasting in an updraft of air, which converts sulphides into oxides and sulphates and serves to make the ore into a physical and chemical form amenable to smelting, as well as removing some volatile impurities. Capacities are normally within the range 1000 to 2500 tons day^{-1}, about 30-35% of the feed comprising lead concentrates, themselves 40-80% lead. This process is a major source of lead emission as well as SO_2, producing 21-260 g of total particulates per kg of lead product, comprising 20-65% lead [1]. The effluent gas stream is at a temperature of approximately 650° C and the particles are fairly large, with 90-99% being >5 μm and 55-85% >40 μm by weight (see Table 5.2).

The sintered ore is then passed to the blast furnace, where it is mixed with coke together with various slags, silica, limestone and recycled lead from emission control devices. In the furnace, the lead is initially converted to oxides by injection of a blast of air. Subsequently the oxides are reduced by carbon and CO to metallic lead. A slag comprising mainly iron and calcium silicates is also formed. The liquid lead flowing from the furnace is about 94-98% pure, and is known as lead bullion.

The characteristics of the exhaust gases from the blast furnace are summarized in Table 5.2. Normally these gases are diluted to 9-15 times their volume during which process a considerable proportion of the CO is oxidized to CO_2. The lead content of these gases comprises very much smaller particles than that from the sinter plant, indicating formation of a fume from condensing lead vapour, as opposed to the dust generated during the sintering process.

An oxidized impurity, or dross, is formed on the surface of the molten bullion, and this is removed as a solid scum. It amounts to 10-35% of the bullion, and contains many metallic impurities. Lead bullion is recovered from the dross by charging a dross reverberatory furnace with 60-70% by weight of dross accompanied by sulphur, coke and a flux. Typically the furnace has a capacity of about 130 tonne day^{-1} of charge. This process causes an atmospheric emission of lead, predominantly in the form of sub-micrometer particles (see Table 5.2).

The latter parts of the lead production process (Fig. 5.1) are those of refining the lead bullion. These are not major sources of lead emissions. Any process in

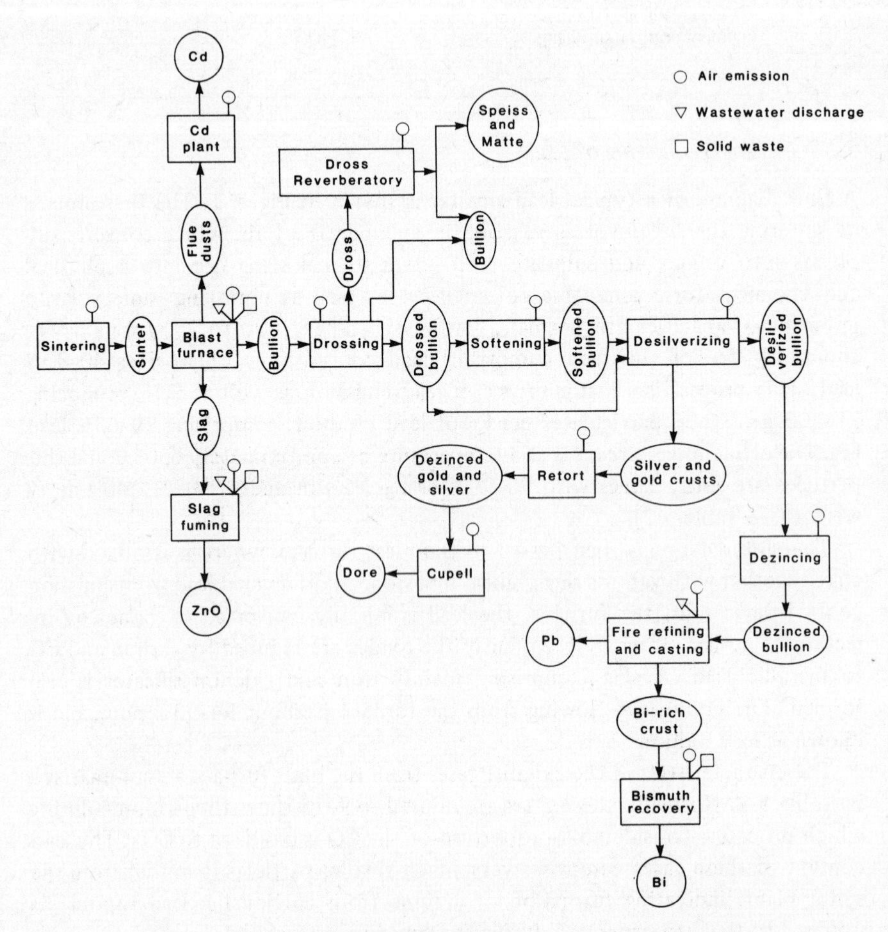

Fig. 5.1 Flow diagram of a typical primary lead smelter (after [1]).

Table 5.2 Typical properties and composition of gas streams from primary lead production processes [1].

Property	Sintering	Blast furnace	Dross reverberatory
Gas flow rate ($m^3 s^{-1}$ (tonne h^{-1} product)$^{-1}$)	1.1	0.63	0.31
Temperature (° C)	650	595–705	760–980
Moisture content (% v/v)	–	<1	Negligible
Particle loading ($g m^{-3}$)	2–57	2–25	0.9–10
Particle size distribution	<20–40 μm; 15–45 wt% <10–20 μm; 9–30 wt % <5–10 μm; 4–19 wt % <5 μm; 1–10 wt %	0.03–0.3 μm (majority)	<1 μm (majority)
Lead content of particulate (% w/w)	20–65	10–40	13–35
SO_2 content (% v/v)	4–6	0.01–0.25	<0.05
CO content (% v/v)	–	25–50	–
Emission factors:			
(i) particulate (g kg^{-1} Pb produced)	21–260	87–125	10
(ii) lead (g kg^{-1} Pb produced)	4.2–170	8.7–50	1.3–3.5

which lead is raised to a temperature in excess of 500° C, however, is a potential source of fume.

Fugitive emissions also arise in the lead production processes, and estimates of their magnitude appear in Table 5.3. These are caused mainly by the sintering operations, lead ore concentrate handling and transfer, and zinc fuming furnace vents [1]. They are mostly of a relatively large particle size and cause environmental contamination within a range of up to about 1 km.

The techniques used in cleansing process gas streams of lead will be described in section 5.2.

Table 5.3 Estimated fugitive dust emissions from operations at a primary lead smelter [1].

Process	Pb (% w/w)	Uncontrolled particulate emissions ($g\ kg^{-1}$)
Ore concentrate storage	37	0.16
Return sinter transfer	19	2.25–6.75
Sinter sizes and storage	58	0.28–1.22
Sinter product dump area	31	0.0025–0.0075
Blast furnace roof vents	47	0.04–0.12
Blast furnace upset	27	3.5–11.5
Lead casting roof ducts	38	0.22–0.66
Zinc fuming furnace area	3	1.15–3.45

5.1.1.2 Primary smelting of other metals

As noted in the last section, lead ores often occur naturally in conjunction with the ores of other metals, and hence primary production of these metals may give rise to emissions of lead. Most important in this respect are the production of zinc and copper.

The lead content of zinc ore concentrates may vary between a few tenths of a per cent and several per cent. Initially the concentrates are roasted to remove sulphur. Subsequently one of two processes is followed, either a pyrometallurgical or electrolytic process. In pyrometallurgical processing, the roasted ore is first sintered, and then reduced to zinc metal in a retort furnace. Both the sintering process and horizontal and vertical retorts are significant sources of lead emissions. The typical compositions of uncontrolled effluent gas streams from these processes appear in Table 5.4. Fugitive emissions also arise from sources similar to those discussed under primary lead smelting, and these will be highly variable between works, dependent upon the process design and the control measures.

In the electrolytic production of zinc, the roasted ore concentrate is leached with sulphuric acid and after purification of the resultant solution, zinc is recovered by electrolysis. No significant atmospheric emissions result from this process, other than those from the roasting process.

Table 5.4 Typical properties and composition of gas streams from primary zinc production processes [1].

Property	Sintering	Horizontal retort	Vertical retort
Gas flow rate ($m^3\ s^{-1}$ (tonne $h^{-1})^{-1}$ product)	1.2–2.7	3.6–5.7	6.4
Temperature (° C)	95–370	–	590
Particle loading ($g\ m^{-3}$)	0.9–10.3	0.1–0.32	2.1
Particle size distribution	<10 μm; 100 wt %	–	<10 μm; 100 wt %
Lead content of particulate (% w/w)	30–55	0–3	4–5
SO_2 content (% v/v)	4.5–7	–	–
CO_2 content (% v/v)	–	12–17	2.5–3.0
Dew point (°C)	50–60	–	–
Emission factors:			
(a) particulate ($g\ kg^{-1}$ product)	45	40	50
(b) lead ($g\ kg^{-1}$ product)	13.5–25	1.2	2–2.5

Production of primary copper involves an initial roasting of the ore concentrate. This is followed by smelting in a reverberatory or electric furnace, the former being more commonly used. The copper matte produced is purified in a converter, forming blister copper which is further refined, mostly by an electrolytic process. The composition of the most contaminated effluent streams appears in Table 5.5. Fugitive emissions inevitably also accompany the production processes.

5.1.1.3 Secondary smelting of lead

A substantial proportion of the lead used at present is recycled lead, produced from lead scrap and residues by secondary smelting. Emission factors for the processes used, and annual US emissions from secondary lead smelting are given in Chapter 2. In the US, about two thirds of the output of the secondary lead industry is produced using blast furnaces or cupolas, although reverberatory and pot furnaces are also used [1]. A typical blast furnace, with pollution control equipment is shown in Fig. 5.2. The furnace is charged from the top, whilst air

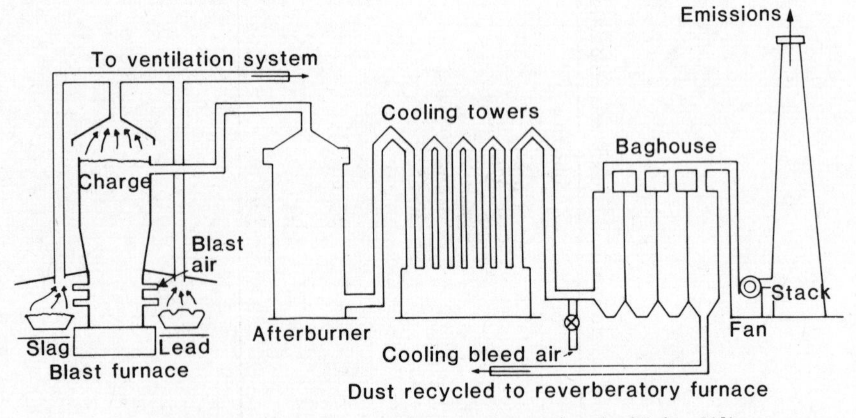

Fig. 5.2 A blast furnace for production of secondary lead (after [1]).

is blown in through tuyeres near the base. The furnace is charged with lead scrap, coke, limestone, scrap iron and re-run slag. Approximately 70% of the molten charge material is tapped off as hard lead (5–12% antimony) [1]. The characteristics of the effluent gases from a secondary lead blast furnace are given in Table 5.6. This ignores fugitive emissions which may be substantial (6 g kg^{-1} of charge for the blast furnace) [1]. Before cooling the effluent stream prior to the baghouse, the gases pass through an afterburner (oil or gas fired) which combusts the hydrocarbons and carbon monoxide present.

5.1.1.4 Refining of lead

The crude lead bullion produced by primary and secondary smelting of lead

Table 5.5 Typical properties and composition of gas streams from primary copper production processes [1].

Property	Roasting	Reverberatory furnace	Copper convertor
Gas flow rate ($m^3\ s^{-1}$ (tonne $h^{-1})^{-1}$ product)	$0.73x + 1.6$*	$0.73x + 1.6$*	0.83–1.0
Temperature (° C)	650	370	1200
Moisture content (%)		18	4.8–6.8
Particle loading ($g\ m^{-3}$)	14–55	5–12	12
Particle size distribution	<10 μm; 15 wt % >10 μm; 85 wt %		<10 μm; 50%
Lead content of particulate (% w/w)	0.5–12	8.3	0.83–8.6
SO_2 content (% v/v)	0.5–2 (multiple hearth) 12–14 (fluid bed)	0.5–2.5	6–7
Emission factors:			
(i) particulate ($g\ kg^{-1}$)	22.5	10 $g\ kg^{-1}$ Cu concentrate	120 $g\ kg^{-1}$ product
(ii) lead ($g\ kg^{-1}$)	$1.2P$†	0.8 $g\ kg^{-1}$ Cu concentrate	1.0–10 $g\ kg^{-1}$ product

*x = tonne h^{-1} production capacity.

†P = Percentage lead in concentrate

Table 5.6 Typical properties and composition of the effluent gas stream from a secondary lead blast furnace [1].

Property	
Gas flow rate ($m^3 s^{-1}$ (tonne h^{-1} product)$^{-1}$)	7.3
Temperature (° C)	730
Moisture content (% v/v)	5
Grain loading ($g m^{-3}$)	9
Particle size distribution	>3μm; 58 wt % 2–3μm; 23 wt % 1–2 μm; 17 wt % 0–1 μm; 2 wt %
Lead content of particulate (% w/w)	23
Emission factors:	
(i) particulate (g kg^{-1} product)	120
(ii) lead (g kg^{-1} product)	28

normally requires refining to remove impurities (mainly copper, arsenic, antimony tin, silver, gold and bismuth) prior to industrial use [1]. Refining may be based upon electrochemical or pyrometallurgical techniques and in the case of lead there is no purely electrochemical method.

The major processes involved in lead refining are: decopperizing, elimination of arsenic, antimony and tin, desilverizing and debismuthizing [2]. Decopperizing is a pyro-refining process (using oxidation–precipitation reactions to form physically separable compounds that can be removed by differences in specific gravity or solubility) [2], whilst the other processes may be carried out either electrolytically or pyrometallurgically. In either case polluted effluent streams are produced, due to volatilization of metal in the pyro-refining processes, and in the pyrometallurgical processing of slimes produced in the electrolytic methods, to recover the metals present. A more detailed account of these processes, and of the gas streams requiring treatment is given by Barbour, Castle and Woods [2].

5.1.1.5 Production of ferrous metals and alloys

The ferrous metals industry is a major contributor of lead emissions to the atmosphere (see emission inventory, Chapter 2). The industrial raw materials are not rich in lead, but as a consequence of the massive size of the industry, and the high process temperatures used, considerable emissions of lead result. In 1975, it is estimated that in the US, 1770 tonnes of lead were emitted to the atmosphere in the production of ferrous metals and alloys: 605 tonnes from iron and steel plants, 1080 tonnes from gray iron foundries and 82 tonnes from ferroalloy production facilities, excluding fugitive emissions [1].

It is beyond the scope of this text to give detailed descriptions of iron and steel-making and foundry processes. The interested reader is referred to a report by the US Environmental Protection Agency [1]. Emission factors for the major processes are listed in Chapter 2. The basis for selection of control techniques is the same as that for the control of other industrial emissions of lead, and the commonly used control methods are listed later in this chapter.

5.1.2 Industrial processes using lead

5.1.2.1 Production of lead oxides and pigments

The most important oxides of lead are litharge (PbO), lead dioxide (PbO_2) and red lead (Pb_3O_4). Black oxide (a mixture of 60–80% PbO with finely divided metallic lead) is used in the manufacture of lead–acid batteries (see Section 5.1.2.2), and in 1975 in the US approximately 70% of lead used in oxides and pigments was used for storage battery manufacture. The estimated emission of lead to the atmosphere from manufacture of the lead oxides and pigments was 112 tonne y^{-1} [1].

Both litharge and black oxide are prepared in processes involving the oxidation of metallic lead. The properties of the exhaust gases from production of black oxide by the ball mill or Barton process are shown in Table 5.7. A fabric filter (or baghouse) is an essential part of plant design and the quoted emission factor is determined after passage through the filter.

Table 5.7 Typical properties and composition of the effluent gas stream from lead oxide ball mill and Barton pot processes [1].

Property	
Gas flow rate ($m^3\ s^{-1}$ (tonne h^{-1} lead charged)$^{-1}$)	1.2
Temperature (° C)	120
Grain loading ($g\ m^{-3}$)	7–11
Particle size distribution*	0–1 μm; 4 wt %
	1–2 μm; 11 wt %
	2–3 μm; 23 wt %
Lead emission factor† ($g\ kg^{-1}$ product)	0.22

*Baghouse catch.
†Emissions after baghouse.

The more important lead pigments are red lead (Pb_3O_4), white lead (mainly $2PbCO_3 \cdot 2Pb(OH)_2$) and lead chromate ($PbCrO_4$). The quantities produced are rather modest and hence any pollution problems are liable to be of a very localized nature.

5.1.2.2 Storage battery manufacture

Manufacture of lead–acid storage batteries is a major use of lead, but only a rather minor contributor to atmospheric pollution.

The plates used in lead–acid batteries are made from moulds cast from a lead alloy which are coated with paste consisting of a mixture of black powder

(PbO and Pb), water, sulphuric acid and organic expander and other constituents dependent upon whether the paste is used for positive or negative plates [1]. The plates are then dried, welded together and placed within a case. The battery is shipped in this dry form and is only subsequently filled with sulphuric acid and charged.

Minor losses of lead occur from all of the individual operations in battery manufacture. A typical uncontrolled emission factor is estimated at 8.0 kg per 1000 batteries, with an overall emission control of 80% within the industry [1].

5.1.2.3 Lead alkyl manufacture

The production of lead alkyls takes the form primarily of tetraethyllead (TEL) and tetramethyllead (TML). Mixed alkyls, such as diethyldimethyllead may be prepared subsequently by reaction of TML and TEL in the presence of a Lewis acid, but this is relatively unimportant.

Both TEL and TML may be manufactured either by alkylation of a sodium-lead alloy, or by the electrolysis of an alkyl Grignard reagent. The first process is by far the most important and this will be described in some detail. Fig. 5.3 shows a flow diagram of the processes involved in TEL manufacture. TML is made by an essentially similar route, with the input of methyl chloride in place of ethyl chloride and the addition of a catalyst such as aluminium chloride, with acetone as a diluent.

Molten lead and molten sodium are combined in a ratio of 9:1 by weight, and the resultant alloy is solidified and flaked. The flaked alloy is passed to an autoclave which is charged with ethyl chloride for a period of 1 h or more. The reaction takes place in the presence of an acetone catalyst at a temperature of 70–75° C and a pressure of 350–420 kPa.

$$4\,NaPb + 4\,C_2H_5Cl \rightarrow (C_2H_5)_4Pb + 4\,NaCl + 3\,Pb$$

After release of the pressure from the autoclave, the contents are discharged to steam stills from which unreacted ethyl chloride is removed by distillation and TEL is steam distilled. TEL is separated from the TEL/water mixture and purified by air blowing and/or washing with dilute oxidizing agents.

The residue from the steam still is rich in lead and is sluiced to a sludge pit to allow recovery of the lead. The atmosphere above the sludge pit is extracted, producing a lead alkyl-rich effluent stream. The sludges are leached with water, the pit sediments are removed, dried and fed to a reverberatory furnace for recovery of lead which is recycled directly to the start of the process.

Particulate lead emissions arise primarily from the lead recovery furnace through which three quarters of the lead is recycled. Uncontrolled, an emission factor of 28 g kg^{-1} of alkyl lead product is typical [1]. Particulate emissions also arise from the lead melting furnace and alloy reactor, but these are negligible relative to the lead recovery furnace.

Emissions of gaseous lead alkyls arise from process vents at a rate of 2 g (Pb)

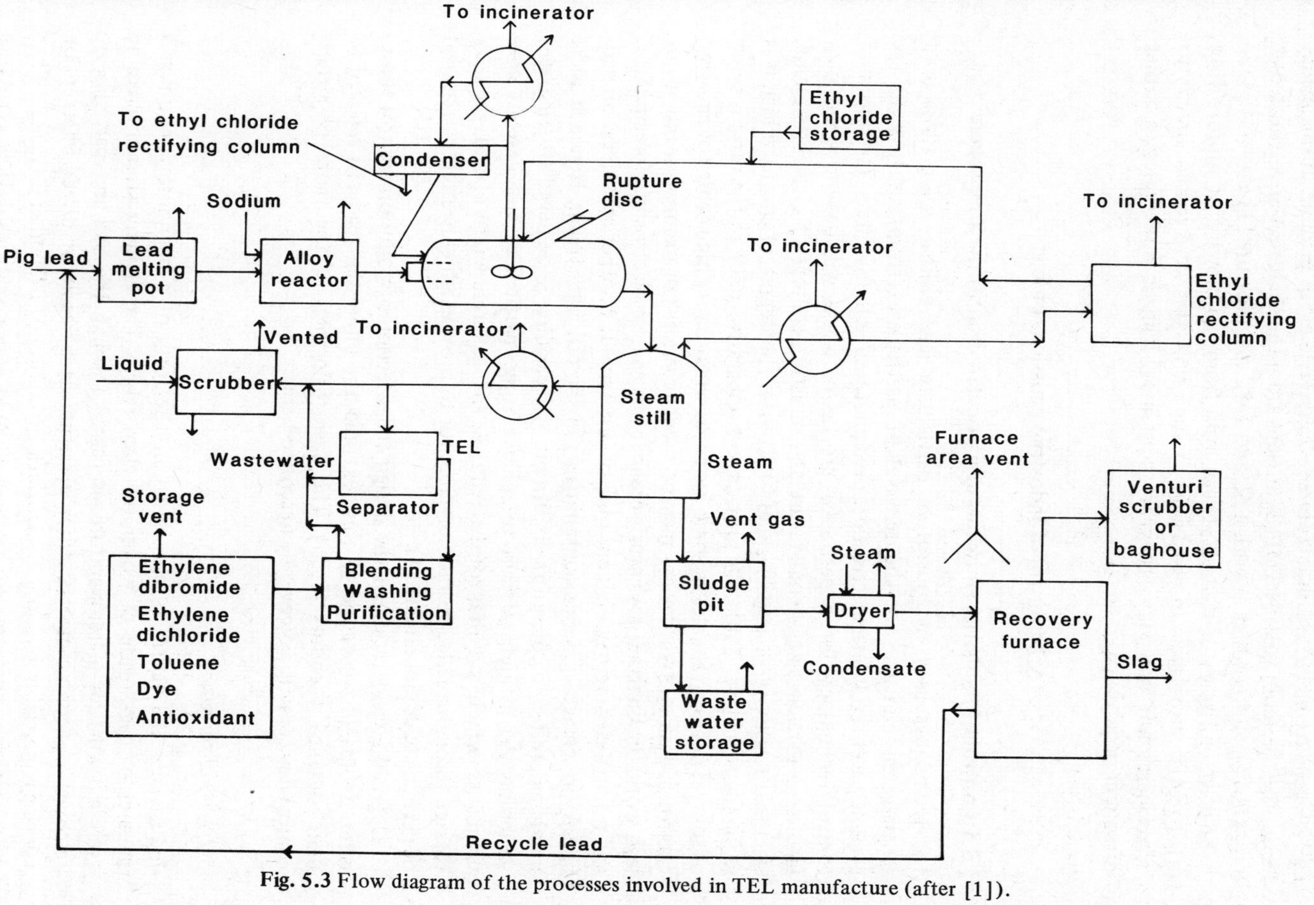

Fig. 5.3 Flow diagram of the processes involved in TEL manufacture (after [1]).

kg^{-1} of product in TEL manufacture and 75 g (Pb) kg^{-1} of product during manufacture of the more volatile TML. Emissions from the sludge pit are estimated at 0.6 g (Pb) kg^{-1} of product in both TEL and TML manufacture [1].

Fugitive emissions of lead alkyls can arise from the blowing of rupture discs fitted to the autoclaves. In years gone by such emissions were vented directly to atmosphere, but nowadays they are contained. This is nonetheless a rather infrequent occurrence.

5.2 Control of industrial emissions of lead

5.2.1 Control of industrial emissions of particulate lead to the atmosphere

The techniques used in control of particulate lead emissions are those used generally in industry for the removal of particulates from gas streams, i.e. cyclones, wet scrubbers, electrostatic precipitators and fabric filters. The optimum control technology for a particular process or works will depend upon such factors as particle sizes, temperature and moisture content of effluent streams, electrical resistivity of particles and, of course, the degree of control which it is required to achieve. Table 5.8 provides a comparative guide to the various types of control device available. In some instances where a very high degree of emission control is desired, two devices may be used in series. For example at one major lead works in England, a Venturi scrubber precedes an electrostatic precipitator, and a most satisfactory effluent quality is obtained. Another consideration may be the presence of gaseous contaminants in the effluent stream. Hence at most municipal refuse incinerators in Britain, gas cleaning is achieved by first wet scrubbing the effluent stream to remove some particulate and gaseous contaminants which simultaneously reduces the gas temperature, and then by passage through an electrostatic precipitator to provide a highly effective removal of particulate contaminants [4].

Detailed consideration of the design and selection of pollution control equipment is outside the scope of this text and the interested reader is referred to more detailed accounts [5, 6]. The basic principles of the more important control devices will, however be described here.

5.2.1.1 Fabric filters

Industrially the fabric filter is used in the form of a baghouse containing a number of bags made of woven or felted fabric. The effluent stream, possibly after a reduction in temperature, is passed through the bag and particles are removed by contacting the filter elements as a result of direct interception (sieving), inertial impaction, electrostatic attraction or Brownian diffusion. When a considerable dust deposit has accumulated and a significant flow resistance is encountered the bag is cleaned by shaking, rapping or causing vibration by an air jet applied from the outside. Accumulated dust is removed at the base in a hopper. The typical form of an industrial baghouse is shown in Fig. 5.4. Baghouses are highly efficient, but can tolerate only rather modest temperatures, and are subject to damage by corrosive gases.

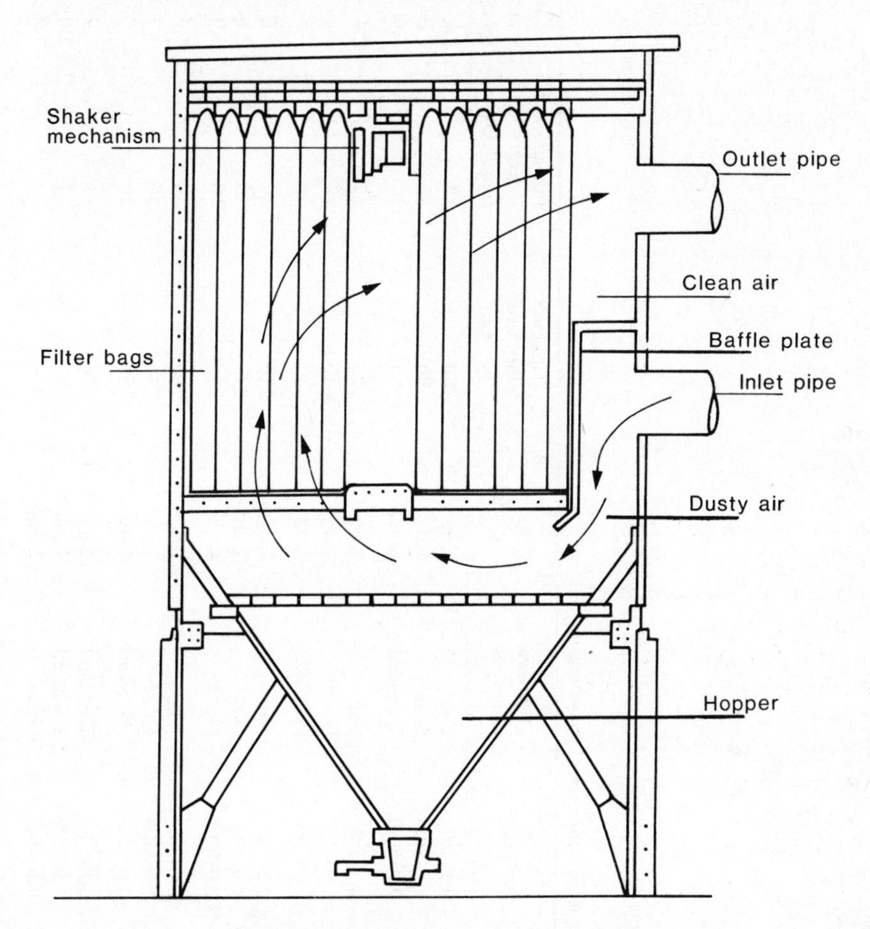

Fig. 5.4 Schematic diagram of a baghouse filter.

5.2.1.2 Electrostatic precipitators

In the electrostatic precipitator the effluent stream passes through a very high electric field established between a network of negatively charged wires and grounded plates or tubes with a potential difference of 30 kV or more (Fig. 5.5). The dust becomes charged by bombardment with negative ions and migrates to the grounded electrode where it is deposited. Removal is by rapping or washing with water. Efficiency is generally high, but is less for small particles, and for materials of high resistivity.

Table 5.8 Comparative guide to particulate control devices (Based upon [3]).

Type of collector	Particle size range (μm)	Removal efficiency	Space required	Maximum temperature (° C)	Pressure drop (cm H_2O)	Relative annual cost (per year m^{-3})*
Bag house (cotton bags)	0.1–1.0	Fair	Large	80	10	2.0
	1.0–10.0	Good	Large	80	10	2.0
	10.0–50.0	Excellent	Large	80	10	2.0
Bag house (Dacron, nylon, Orlon)	0.1–1.0	Fair	Large	120	12	2.4
	1.0–10.0	Good	Large	120	12	2.4
	10.0–50.0	Excellent	Large	120	12	2.4
Bag house (glass fibre)	0.1–1.0	Fair	Large	290	10	3.0
	1.0–10.0	Good	Large	290	10	3.0
	10.0–50.0	Good	Large	290	10	3.0
Bag house (Teflon)	0.1–1.0	Fair	Large	260	20	3.3
	1.0–10.0	Good	Large	260	20	3.3
	10.0–50.0	Excellent	Large	260	20	3.3
Electrostatic precipitator	0.1–1.0	Excellent	Large	400	1	3.0
	1.0–10.0	Excellent	Large	400	1	3.0
	10.0–50.0	Good	Large	400	1	3.0

Standard cyclone	0.1–1.0	Poor	Large	400	5	1.0
	1.0–10.0	Poor	Large	400	5	1.0
	10.0–50.0	Good	Large	400	5	1.0
High-efficiency cyclone	0.1–1.0	Poor	Moderate	400	12	1.6
	1.0–10.0	Fair	Moderate	400	12	1.6
	10.0–50.0	Good	Moderate	400	12	1.6
Spray tower	0.1–1.0	Fair	Large	540	5	3.6
	1.0–10.0	Good	Large	540	5	3.6
	10.0–50.0	Good	Large	540	5	3.6
Impingement scrubber	0.1–1.0	Fair	Moderate	540	10	3.3
	1.0–10.0	Good	Moderate	540	10	3.3
	10.0–50.0	Good	Moderate	540	10	3.3
Venturi scrubber	0.1–1.0	Good	Small	540	88	8.0
	1.0–10.0	Excellent	Small	540	88	8.0
	10.0–50.0	Excellent	Small	540	88	8.0

*Includes: Water and power cost, maintenance cost, operating cost, capital and insurance costs.

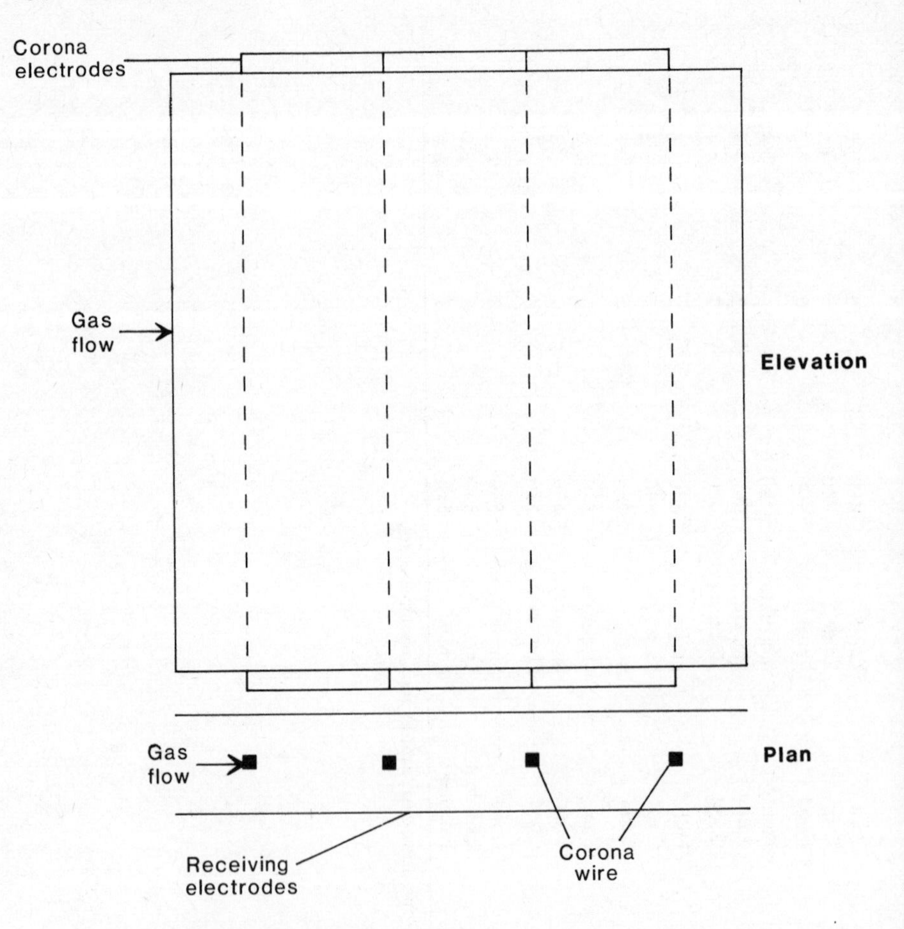

Fig. 5.5 Schematic diagram of an electrostatic precipitator.

5.2.1.3 Inertial collectors

The most commonly used inertial collector is the cyclone (Fig. 5.6). This consists of a cylindrical vessel into which the gas stream is introduced tangentially. The rotational motion of the gas stream causes movement of the particles to the outer wall where deposition and aggregation occur. The dust falls to the base of the cyclone and is removed into a hopper, whilst the cleaned gas stream is removed at the top. The main disadvantage of the cyclone, and of other inertial collectors, is the low efficiency at small particle sizes, and for this reason the cyclone is most commonly used to pre-clean an air stream which is subsequently 'polished' using a more efficient dust arresting device.

5.2.1.4 Wet scrubbers

These come in many shapes and forms and may be used for removal of both

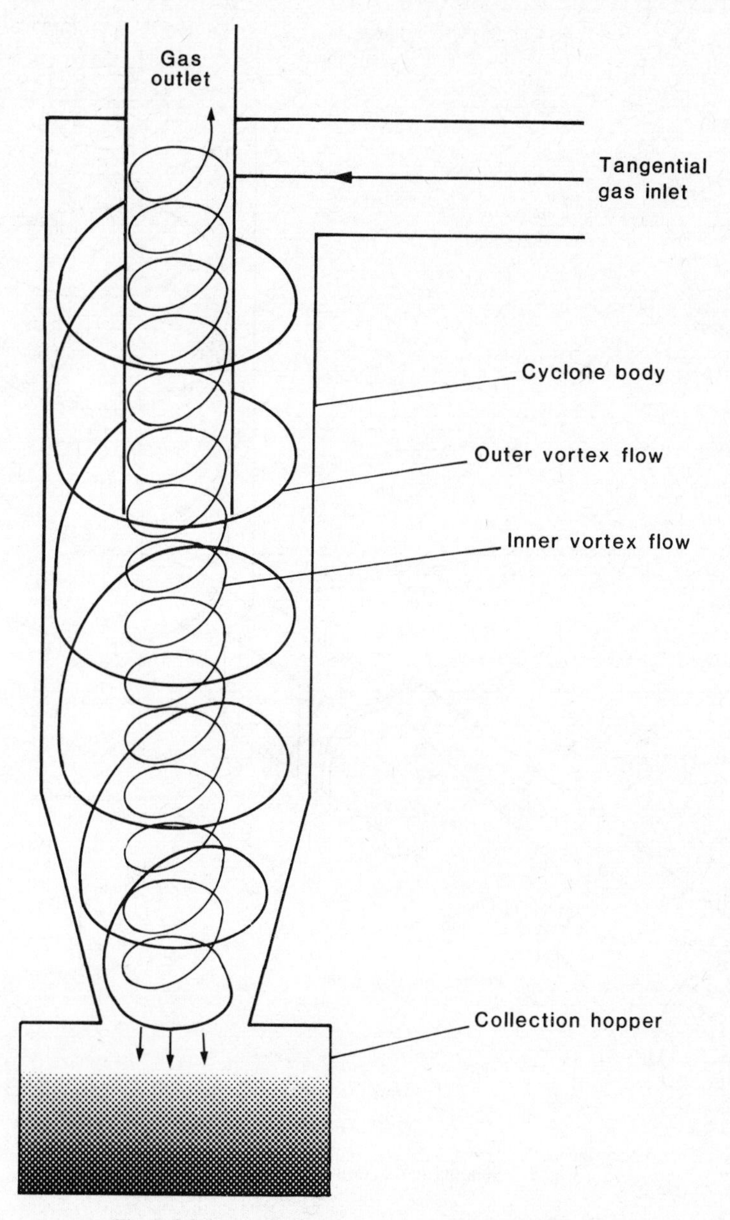

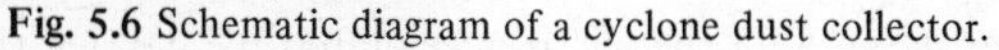

Fig. 5.6 Schematic diagram of a cyclone dust collector.

gaseous and particulate pollutants from an effluent stream. In their simplest form of a spray tower, the wet scrubber involves subjecting the contaminated gas stream to a continuous fine spray of water droplets, which wet and thence remove particulate contaminants. In the most efficient form of scrubber, the

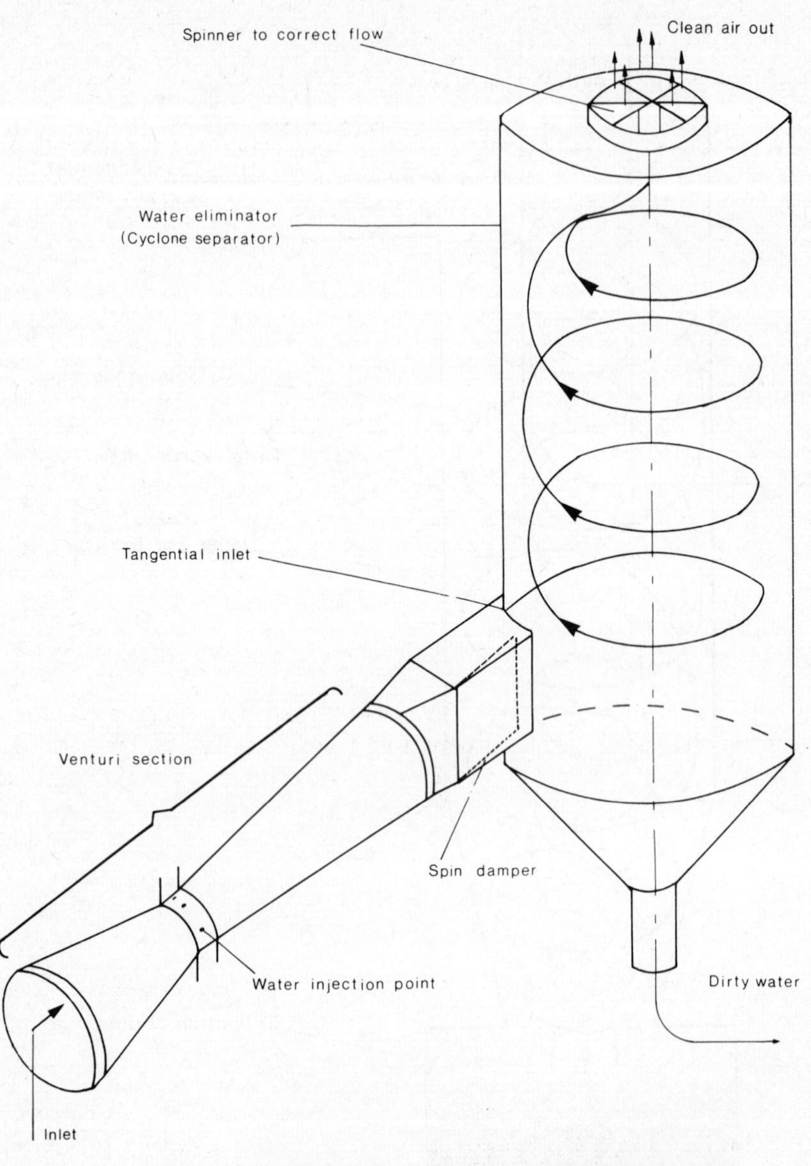

Fig. 5.7 Schematic diagram of a Venturi scrubber.

Venturi scrubber, (Fig. 5.7) the gas is accelerated by passage through a narrow throat into which water sprays are injected. As a result of the high relative velocity of the gas and water droplets and the high turbulence in the exit of the throat, efficiency is high, but the pressure drop is also considerable giving rise to relatively high operational costs. Residual water spray is removed by a cyclone separator (Fig. 5.7).

5.2.1.5 Control systems used in the lead industry

Table 5.9 lists the principal processes from which lead emissions arise, and also indicates the most commonly used control techniques.

It must be remembered that many of the available pollution control devices produce large volumes of contaminated water (e.g. from the Venturi scrubber). These provide a further contaminated effluent stream requiring the application of control technology (see Chapter 6).

5.2.2 Control of industrial emissions of organic lead to the atmosphere

It has been noted (Section 5.1.2.3) that production of tetraalkyllead compounds gives rise to effluent streams contaminated with gaseous organic lead compounds. These cannot be controlled by the application of particulate control technology, and considerable effort has been expended in recent years to achieve satisfactory control. Present practice in England involves the use of an activated carbon absorption plant, followed by recovery of absorbed lead by steam stripping [7].

5.2.3 Chimney heights and dispersion of pollution from a point source

It must not be forgotten that a prime aim of reduction of lead emissions is to limit the exposure of the general population by reducing ambient concentrations of lead around the works. According to the Gaussian plume model, the ground-level concentration of a pollutant emitted from a point source (χ) is given by

$$\chi(x) = \frac{Q}{\pi \sigma_y \sigma_z \bar{u}} \exp\left(\frac{-H^2}{2\sigma_z^2}\right)$$

where the concentration is measured on the plume centre-line, Q is the source strength, $\bar{u}$ is the mean wind velocity at the height of emission, σ_y is the standard deviation of the pollutant concentration distribution in the cross-wind plane, σ_z is the standard deviation of the pollutant concentration distribution in the vertical plane and H is the effective chimney height. The parameters σ_y and σ_z are functions of the downwind distance (x) and the atmospheric stability (degree of turbulence). Hence there is no control over u, σ_y or σ_z. Supposing Q has been reduced to the lowest practicable level by the application of control equipment, then the only variable is H.

The effective chimney height, H, is made up of two components. It is the sum of the actual chimney height and the plume rise, the latter being the additional height to which the plume centre-line ascends due to its buoyancy and efflux velocity. Examination of curves in which the maximum ground-level concentration is plotted as a function of H and atmospheric stability show that χ is

Table 5.9 Summary of emission factors and control techniques commonly used in industry [1].

Industry and process	Uncontrolled lead emission factor		United States 1975 lead emissions		Control techniques[b]		
	($g\ kg^{-1}$)[a]	($lb\ ton^{-1}$)	(tonne)	(US tons)	FF	WC	ESP
Gasoline combustion	$0.20x$[c]	$1.65y$[d]	127 800	140 900	R	R	R
Coal combustion	$0.80L$[e]	$1.6L$[e]	228	257	R	O	T
Oil combustion	$0.5P$[f]	$4.2P$[f]			R	R	R
utilities			45	50	R	R	R
industrial			14	15	R	R	R
other sources			41	45	R	R	R
Solid waste incineration	0.2	0.4	1 170	1 296	O	T	T
Waste oil combustion	$4.8M$[g]	$40M$[g]	5 000	5 480	R	R	R
Lead alkyl manufacturing							
Sodium–lead alloy process			1 000	1 100			
recovery furnace	28	55			T	O	R
process vents, TEL	2	4			R	O	R
process vents, TML	75	150			R	O	R
sludge pits	0.6	1.2			R	O	R
Electrolytic process	0.5	1.0	14	15	R	T	R
Storage battery manufacturing	8.0	17.7[h]	82	90			
grid casting	0.40	0.90			O	O	R
paste mixing	0.50	1.20			O	T	R
three-process	6.60	14.7			T	R	R
lead reclaim	0.35	0.77			O	T	R
lead oxide production	0.05	0.12			T	T	R
Ore crushing and grinding	0.006–0.15	0.012–0.3	493	544	R	R	R
Primary copper smelting							
roasting	$1.2P$[i]	$2.3P$[i]	107	112	R	T	T
reverberatory furnace	0.83	1.7	222	244	R	R	T
converting	1.3	2.6	987	1 085	R	R	T

Primary lead smelting			400	440			
sintering	4.2–170	8.4–340			T	T	O
blast furnace	8.7–50	17.5–100			T	T	R
dross reverberatory	1.3–3.5	2.6–7			T	R	R
Primary zinc smelting							
sintering	13.5–25	27–50	84	93	R	R	T
horizontal retorts	1.2	2.4	17	19	R	R	R
vertical retorts	2–2.5	4–5	11	12	T	R	R
Secondary lead smelting			755	830			
blast furnaces	28	56			T	T	R
reverberatory furnaces	27	53			T	T	R
Brass and bronze			47	52			
reverberatory furnace					T	O	R
–high lead alloys	25	50					
–red and yellow brass	6.6	13.2					
–other alloys	2.5	5.0					
Gray iron production							
cupola furnace	0.3	0.6	950	1 050	O	T	R
reverberatory furnace	0.035	0.07	33	36	O	T	R
electric furnace	0.026	0.05	96	106	R	R	R
Ferroalloy production (electric arc)			31	33			
FeMn	0.055	0.11			T	T	R
FeSi	0.15	0.29			T	T	R
SiMn	0.17	0.34			T	T	O
HCFeCr	0.04	0.08					
Iron and steel production							
sintering	0.0067[j]	0.013[j]	18	20	R	T	O
coking	0.0018[k]	0.0035[k]	11	12	R	O	R
blast furnace	0.062	0.12	91	100	R	T	T
open hearth	0.07	0.14	128	141	R	R	T
basic oxygen	0.10	0.20	130	144	R	T	T

Table 5.9 (*cont.*)

Industry and process	Uncontrolled lead emission factor		United States 1975 lead emissions		Control techniques[b]		
	($g\ kg^{-1}$)[a]	($lb\ ton^{-1}$)	(tonnes)	(US tons)	FF	WC	ESP
electric furnace	0.11	0.22	227	250	T	R	R
Lead oxide production	0.22[l]	0.44[l]	100	110	T	R	R
Red lead production	0.5[l]	0.9[l]	8	9	T	R	R
White lead production	0.28[l]	0.55[l]	0.9	1	T	R	R
Chrome pigments production	0.065	0.13	3.0	3.3	T	R	R
Type metal operations	0.13[m]	0.25[m]	436	480	T	T	O
Can soldering operations	0.16[n]	0.18[n]	60	67	R	R	R
Cable covering operations	0.25[m]	0.5[m]	113	125	R	R	R
Portland cement production							
Dry process							
kilns	0.06	0.11	135	149	T	R	T
coolers and grinders	0.02	0.04	53	58	T	R	O
Wet process							
kilns	0.05	0.10	110	120	T	R	T
coolers and grinders	0.01	0.02	15	17	T	R	O
Metallic lead products	0.75[m]	1.5[m]	77	85	R	R	R
Ammunition manufacturing	0.5[p]	1.0[p]	negligible		R	R	R
Lead glass production	2.5	5.0	56	62	T	R	O
Total 1975 lead emissions			141 380	155 900			

[a] Units are g kg^{-1} produced unless indicated otherwise by footnote.

[b] FF = fabric filter, WC = wet collector, ESP = electrostatic precipitator.
R = rare or never, O = occasional, T = typical.

[c] x = lead content in g cm^{-3}. Resulting emission factor units are kg m^{-3} gasoline. Average 1975 lead content was 0.45 g cm^{-3}.

[d] y = lead content in g gal^{-1}. Resulting emission factor units are lb $(10^3\ gal)^{-1}$ gasoline. Average 1975 lead content was 1.7 g gal^{-1}.

[e] L = lead content of coal in ppm by weight. Resulting emission factor units are in g $tonne^{-1}$ of coal, lb $(10^3\ ton)^{-1}$. US coals average about 8.3 ppm lead.

[f] P = lead content of oil in ppm by weight. Resulting emission factor units are g m^{-3} oil, lb $(10^6\ gal)^{-1}$.

[g] M = lead content of waste oil in percent by weight. (Generally around one per cent). Resulting emission factor units are kg m^{-3} oil, lb $(10^3\ gal)^{-1}$.

[h] Units are kg $(10^3\ batteries)^{-1}$ produced, lb $(10^3\ batteries)^{-1}$ for all processes in battery manufacturing.

[i] p = lead content in copper concentrate in per cent by weight. Average lead content for US concentrates is 0.3 per cent. Emission factor units for all copper operations are expressed in g kg^{-1} concentrate (lb ton^{-1}).

[j] Units are in g kg^{-1} of sinter produced (lb ton^{-1}).

[k] Units are in g kg^{-1} of coal consumed (lb ton^{-1}).

[l] Emission factor given is after control with cyclone/fabric filter product recovery system.

[m] Units are g kg^{-1} of lead processed (lb ton^{-1}).

[n] Units are kg $(10^6\ baseboxes)^{-1}$, lb $(10^6\ baseboxes)^{-1}$.

[p] Units are g $tonne^{-1}$ lead processessed, lb $(10^3\ ton)^{-1}$.

critically dependent upon H, and that a doubling of stack height may reduce the maximum ground-level concentration by nearly one order of magnitude [8, 9]. Hence control measures for industrial lead cannot be instituted without consideration of the optimum chimney height to achieve the required dispersion.

5.3 Legislative regulations affecting industrial lead emissions

5.3.1 United Kingdom

In the UK all major works using lead come within the jurisdiction of HM Alkali and Clean Air Inspectorate [10]. The Inspectorate's policy of enforcing 'best practicable means' of pollution control is well known. Presumptive limits applying to all gases and exhaust streams from scheduled lead processes have been set. These are as follows:

Works category	Gas volume ($m^3\ min^{-1}$)	Individual stack concentration* ($g\ m^{-3}$)	Total mass emission* ($kg\ h^{-1}$)
Class I	200	0.115	0.27
Class II	200–4000	0.023	2.7
Class III	>4000	0.0115	5.4

*Measured as elemental lead.

It is of interest to note that during 1976 the arithmetic mean lead content of lead works emissions examined by HM Alkali Inspectorate was 0.011 $g\ m^{-3}$. It was also noted in the report of the Inspectorate [7] that at the Avonmouth primary lead-zinc-cadmium smelter of Commonwealth Smelting Ltd. (with a capacity of 40 000 tonne of lead annually), emission of lead was about 4 $kg\ h^{-1}$ in 1976, compared with a consent limit of 5.4 $kg\ h^{-1}$.

5.3.2 United States

The approach to air pollution control in the US is very different to that in the UK. Rather than controlling lead via an effluent quality standard as in Britain, an ambient air quality standard is applied. The standard, set recently [11] applies to both industrial and vehicular sources of lead and is described in more detail in Section 5.5. Under this standard it is necessary for industrial sources to control emissions, and if necessary raise chimney heights to limit concentrations of lead in ambient air within the vicinity.

5.4 Control of lead emissions from motor vehicles

Motor vehicles comprise by far the major source of atmospheric lead (Table 2.1). Consequently considerable effort has been applied to the control of lead emissions from them. There are two main approaches.

5.4.1 Reduction of the lead content of gasoline

In most developed countries the lead content of gasoline has been reduced over recent years by statutory control. Hence in the UK the maximum permissible lead content of gasoline was 0.84 g dm^{-3} in 1972, it stood at 0.45 g dm^{-3} in 1978 and is due to be reduced to 0.40 g dm^{-3} in 1981. Alone within the EEC, West Germany has reduced the lead content of gasoline to 0.15 g dm^{-3}.

The picture is not as simple as it might appear at first sight. Lead cannot be removed from gasoline without incurring some penalties. The TML and TEL are added to improve the combustion properties of the gasoline and reduce the tendency for 'knock', or pre-ignition of the fuel–air mixture on the compression stroke. 'Knock' or 'pinking', as it is also known, is damaging to the engine and reduces the power output. Addition of lead increases the anti-knock characteristics of the fuel (expressed as octane number); typically a regular grade fuel without TEL having a research octane number of 86 may be raised to an octane number of 93 by the addition of 0.5 g(Pb) dm^{-3} of TEL.

There are four approaches to the problem of low-lead or lead-free petrol. Firstly, it is possible to accept the use of low octane fuel and to re-design engines accordingly. This entails the use of engines of lower compression ratio, which are generally of lower thermal efficiency and consume more fuel [12]. The second approach is to refine crude oil in such a way as to provide a gasoline composition having a high octane number without addition of lead. This approach is criticized on the grounds of (i) the costs of capital investment in new refinery equipment, (ii) the need to use more crude oil per gallon of gasoline, as further processing is required and (iii) the fact that the high octane lead-free gasoline contains a higher proportion of aromatic compounds. It appears that this may lead to increased emission of the carcinogenic polynuclear aromatic hydrocarbons [13]. The third approach is to change to an entirely different fuel, but this is a long-term approach as yet awaiting the availability of a suitable fuel. Fourthly, there has been intensive research into the possibilities of other, non-lead gasoline additives. The only compound presently in use is methylcyclopentadienyl-manganese tricarbonyl (MMT), but this can only be added in small amounts, giving only a modest improvement in octane rating. Its use has been limited to the US, and has been rather sporadic. Problems include the rather unknown environmental properties of emitted manganese, and the problems of plugged exhaust catalysts.

In the US lead-free gasoline has been widely available since 1974. It must, by law, be used in all new cars of 1975 or later model year and is required to prevent poisoning of oxidation catalysts fitted to the exhaust systems of those cars to meet Environmental Protection Agency emission standards for carbon monoxide and hydrocarbons. More recently, measures have been introduced to phase down the lead content of gasoline as a pre-requisite to the achievement of the ambient air quality standard for lead (Section 5.5).

5.4.2 Removal of particulate lead from vehicle exhaust gases

This is the second major approach to limiting vehicle-emitted lead. Various devices have been produced, each capable of substantially reducing the lead content of exhaust gases. These vary in character from being similar to a conventional silencer (muffler) packed with γ-alumina-coated steel wool, to the du Pont lead trap (Fig. 5.8) in which the exhaust gases first pass through a bed of alumina pellets, and then through two cyclones before discharge.

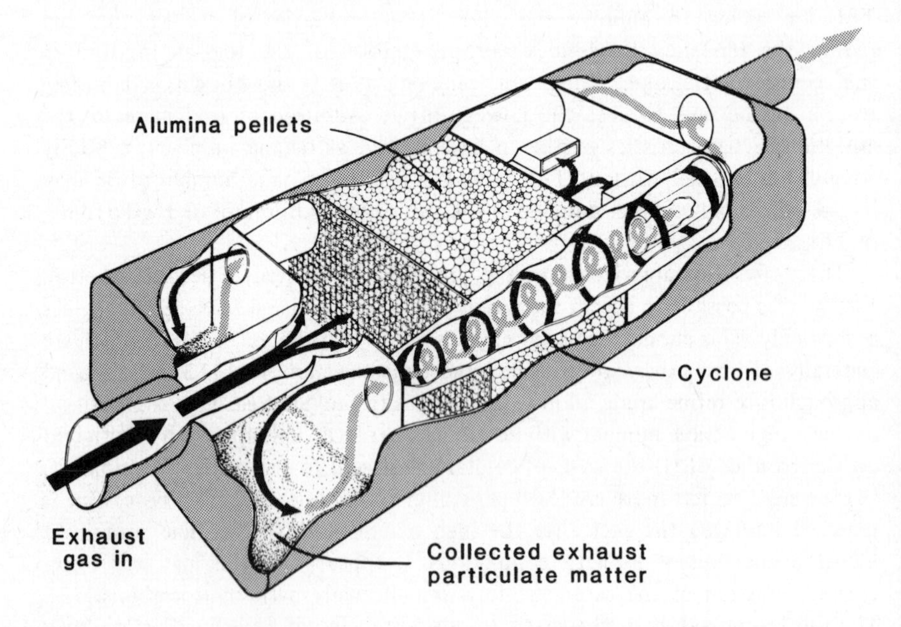

Fig. 5.8 The du Pont exhaust lead trap [14].

A demonstration of the effectiveness of the du Pont trap appears in Fig. 5.9, in which the emissions from cars fitted with a standard exhaust and with the lead trap are compared. High efficiency of particle trapping is achieved, especially for the larger particle sizes (Table 5.10). Work at the Warren Spring Laboratory, England, showed that the lead emitted with standard exhausts and lead traps was respectively 70.0 and 37.5% of the lead input to the engine over a prolonged road test [15].

The cost of the lead trap is relatively modest, and re-cycling of the collected lead is an economic possibility. No country has yet made fitting of the lead trap compulsory.

5.4.3 Lead pollution from waste oil disposal

Around 10% of the lead input to a car ends up in the engine oil. Although this

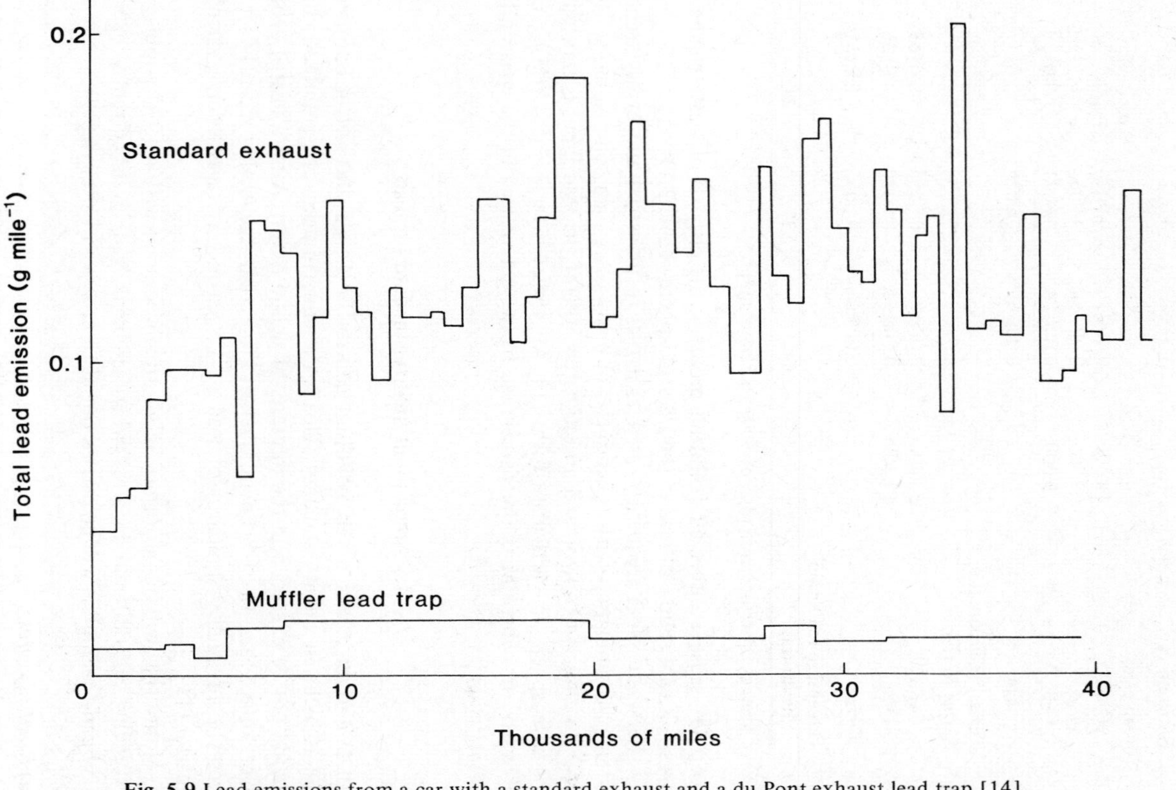

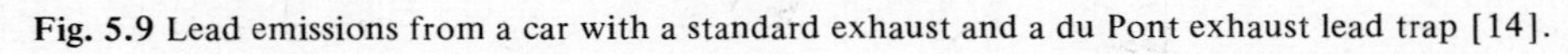

Fig. 5.9 Lead emissions from a car with a standard exhaust and a du Pont exhaust lead trap [14].

Table 5.10 Lead trap performance on 1970 Chevrolets [14].

Vehicle	Mileage	Lead emission rate (g $mile^{-1}$) for particles sizes (μm) >9	1–9	<1.0
Standard car*		0.038	0.023	0.047
Trap cars				
C-82	14 500	0.000 76	0.0024	0.0177
C-83	3 300	0.000 56	0.0031	0.0150
C-85	9 000	0.000 74	0.0021	0.0130
Average		0.000 69	0.0025	0.0147
Percentage reduction due to trap		98%	89%	69%
C-82	53 600	0.000 70	0.0053	0.0165
C-85	51 200	0.000 51	0.0029	0.0136
Average		0.000 60	0.0041	0.0151
Percentage reduction due to trap		98%	82%	68%

*Average of results from tests of 5000 to 55 000 miles

does not cause a direct air pollution problem, the disposal of waste engine oil is a significant source of lead in the US atmosphere (Table 2.1).

The waste oil is frequently used as a fuel in industrial and utility boilers, often in combination with other fuels. In many cases such combustion sources have no emission control, and an estimated 50% of the lead input is emitted to the atmosphere in particulate form [1]. An emission factor of 4.8(M)kg m^{-3} of waste oil, where M is the percentage by weight of lead in the waste oil, is estimated [1].

5.5 Ambient air quality standard for lead

An important aspect of air pollution control in the US, but not in Europe, is the ambient air quality standard. This is the concentration of a pollutant which should not be exceeded in the interests of public health. After a heated debate over some years, the US Environmental Protection Agency has recently set an ambient air quality standard for lead of 1.5 μg m^{-3} averaged over a calendar quarter [11].

Monitoring will be performed in all urbanized areas with a 1970 population exceeding 500 000 and in areas where ambient lead levels currently exceed 1.5 μg m^{-3}. Sampling will be carried out at three types of site [16]:

(a) *Roadway sites*
These are located adjacent to major highways, and placed between 5 and 15 m from the closest traffic lane. Monitors will be sited close to residences and not more than 5 m above ground level.

(b) *Neighbourhood sites*
These are located in an area of high traffic density and high population density, but not necessarily adjacent to a major road. The monitors will be sited at least 15 m from the nearest road with more than 2000–3000 vehicles day^{-1}. Monitors should be not more than 5 m above ground level. It is desirable for monitors to be placed in areas which children frequent, and for siting to expose the monitor to both mobile and stationary sources of lead.

(c) *Street canyon sites*
These are located in areas of high traffic and population density. Monitors will be sited in city centre streets of high traffic density lined with high buildings and should be sited at no more than 5 m above ground level.

The minimum acceptable sampling frequency for a quarterly average is one 24 h sample every six days [11].

Particular attention is to be paid to sites in the vicinity of point sources of lead, i.e. primary lead smelters, secondary lead smelters, primary copper smelters, lead alkyl manufacturing plants, lead–acid battery plants producing more than 2000 batteries per day, as well as any other stationary source emitting 25 or more tons of lead per year [11].

5.6 Control of lead within the workplace

There are a number of industries where the use of lead gives rise to substantial exposure of individuals in the workplace. Such exposure may be via inhalation, ingestion or skin absorption, the latter being important only in the case of lead alkyls.

In a recent document, the UK Health and Safety Commission defines two classes of lead worker: those whose exposure is considered significant and those whose is not [17]. Exposure is considered significant if workers are exposed to (a) levels of airborne lead which are liable to be in excess of one half of the industrial hygiene standard (see later), (b) a substantial risk of ingesting lead or (c) any risk of skin contact with concentrated lead alkyls. Work which is regarded as significant appears in Table 5.11 and that which is insignificant is in Table 5.12.

In all cases, it is proposed that employers will be required to provide information, instruction and training to employees and to fulfil a considerable range of requirements relating to the provision of process controls. They will also have to provide respiratory protective equipment if industrial hygiene standards are exceeded, protective clothing, works cleansing and washing facilities for the workforce. If lead exposure is considered significant, then both lead-in-air monitoring and medical surveillance of the workforce is required.

5.6.1 Industrial hygiene standards for lead

The threshold limit values (TLV) in the UK for a time-weighted average (TWA) (TLV–TWA) over an 8 h shift are as follows:

Table 5.11 Work regarded as significant in terms of lead exposure in the UK [17].

Types of lead work where there is liable to be significant exposure to lead (unless adequate controls are provided)	Examples of industries and processes where such work could be carried out
Lead dust and fumes	
1. High temperature lead work (above 500° C) E.g. lead smelting, melting, refining, casting and recovery processes	–lead smelting and refining; non-ferrous metals, e.g. gun metal, leaded steels manufacture; scrap metal and wire-patenting processes.
Lead burning, welding and cutting	–of lead coated and painted plant and surfaces in demolition work; shipbuilding, breaking and repairing; chemical industry; miscellaneous industries.
2. Work with lead compounds which gives rise to lead dust in air. E.g. any work activity involving a wide variety of lead compounds (other than low solubility lead compounds).	–manufacture of lead batteries, paints and colours, lead compounds, rubber products, glass, pottery, manufacture and use of lead arsenate as pesticide (agriculture and allied work). –manufacture of detonators (explosives industry).
3. Abrasion of lead giving rise to lead dust in air E.g. dry discing, grinding, cutting by power tools	–miscellaneous industries e.g. motor vehicle body manufacture and repair of leaded car bodies, firing of small fire arms.
4. Spraying of lead paint and lead compounds Other than paint conforming to BS 4310/68 and low solubility lead compounds	–painting of bridges, buildings etc. with lead paint.
Lead alkyls	
1. Production of concentrated lead alkyls	–lead alkyl manufacture.
2. Blending of lead alkyls into gasolines	–blending processes at oil refineries.
3. Entry into fixed or mobile plant and vessels, e.g. road tankers, rail tankers or sea tankers carrying leaded gasoline, e.g. for inspection, cleaning and maintenance purposes.	–plant and vessel entry at oil refineries, at transport terminals and any place where such work is carried out.

Table 5.12 Work which is not normally significant in regard of lead exposure in the UK [17].

Types of lead work where there is not liable to be significant exposure to lead	Examples of industries and processes where such work could be carried out
Lead dust and fumes	
1. Work with galena (lead sulphide)	–mining and working of galena when its character or composition is not changed.
2. Low temperature melting of lead (below 500° C) (Such low temperatures control the fume but some care is still required in controlling any dust from dross)	–plumbing, soldering, linotype and monotype casting processes in printing industry.
3. Work with low solubility inorganic lead compounds when tested for low solubility	–painting with low solubility paints conforming to BS 4310/68.
4. Work with materials which contain less than 1% total lead	
5. Work with lead in emulsion, granular or paste form where the moisture content is such and is maintained so that lead dust and fume cannot be given off throughout the work duration	–the use of oil bound lead paints and some stabilizers for plastics.
6. Handling of clean solid metallic lead e.g. lead ingots, pipes, sheets etc. where there is a slight risk of oxide surface contamination	–miscellaneous metal industries, metal stock-holding, general plumbing with sheet lead.
7. Lead emissions from petrol driven vehicles (other products of combustion such as carbon monoxide are the major risk)	–testing of petrol driven engines e.g. motor car testing.
Lead alkyls	
1. Any exposure to lead alkyl vapours from leaded gasoline where the lead content is limited under the Motor Fuel (Lead Content of Petrol) Regulations	–work with such leaded gasoline including e.g. the filling of petrol driven vehicles on garage forecourts.

For inorganic lead dust and fume in air 0.15 $mg(Pb)m^{-3}$
For tetraethyllead in air 0.10 $mg(Pb)m^{-3}$
For tetramethyllead in air 0.15 $mg(Pb)m^{-3}$

TLV short-term exposure limits have also been recommended. These are the maximal concentrations to which workers may be exposed for up to 15 min continuously, with at least 60 min between exposure periods, given that the TLV-TWA is not exceeded.

These are:

For inorganic lead dust and fume in air	0.45 mg(Pb) m^{-3}
For tetraethyllead in air	0.30 mg(Pb) m^{-3}
For tetramethyllead in air	0.45 mg(Pb) m^{-3}.

The above standards are presently applicable in most countries of the world, although in the US it is currently proposed to reduce the TLV-TWA for inorganic lead to 0.10 mg(Pb) m^{-3} [11].

5.6.2 Monitoring for industrial hygiene

In situations where lead exposure is considered significant, lead-in-air monitoring is carried out on a routine basis. This takes the form of both personal monitoring, in which samplers are attached to some individual workers with the inlet attached to the lapel, as well as monitoring at representative points within the works atmosphere.

Medical surveillance of exposed workers differs according to whether exposure is to inorganic lead or lead alkyls. In the case of inorganic lead the appropriate measure of recent exposure is the blood lead level (see Chapter 7). This is normally determined every three months and any worker whose blood lead exceeds 80 μg (100 ml)$^{-1}$ of whole blood is removed from lead exposure at work. Determinations of haemoglobin concentrations and clinical assessments are also used to evaluate the health of lead workers. For workers exposed to concentrated lead alkyls a measure of lead in urine is made at 6 weekly or more frequent intervals, and blood leads are measured annually. Any worker with a urinary lead value in excess of 0.8 μmol dm^{-3} is suspended from lead work until considered medically fit to return.

5.6.3 Control of lead exposure of workers

There are several methods of controlling lead in the workplace. Some are specific to individual industries, others are applied generally where lead exposure occurs. These can be viewed under two headings:

(a) *Plant design*
Where possible all processes in which dust or fume is generated are enclosed and vented to the outside. Where this is not possible the workplace atmosphere is forcibly extracted to provide ample ventilation and to keep lead-in-air to acceptable levels. In areas where industrial hygiene standards are liable to be exceeded, personal respirators are issued to workers to provide a constant supply of uncontaminated air for breathing. Deposited dusts are frequently removed.

(b) *Personal hygiene of workers*
This is an equally important element of exposure control. Workers are educated

in the risks of lead exposure and shown how poor personal hygiene leads to increased exposure. Provision of washing facilities, protective clothing and the prohibition of eating, drinking and smoking in the workplace are all facets of industrial hygiene practice. In some cases precautions have been found necessary to prevent workers going home in contaminated clothing, causing consequent contamination of the home and increased exposure of the family.

Detailed consideration of the measures appropriate to a particular industry is beyond the scope of this book, and the interested reader is referred to recent publications of the Health and Safety Commission [10, 17].

References

[1] US Environmental Protection Agency (1977), *Control Techniques for Lead Air Emissions*, EPA–450/2–77–012.

[2] Barbour, A. K., Castle, J. F. and Woods, S. E. (1978), Production of non-ferrous metals, in *Industrial Air Pollution Handbook*, (ed. A. Parker), McGraw-Hill, New York, pp. 419–93.

[3] Stern, A. C., Wohlers, H. C., Boubel, R. W. and Lowry, W. P. (1973), *Fundamentals of Air Pollution*, Academic Press, New York and London, pp. 403–26.

[4] Perry, R. and Harrison, R. M. (1975), General Trends in Air Pollution and Some Aspects of its Relation to Water Pollution Control, *Public Health Engineer*, 18, 158–68.

[5] Darby, K. (1978), Removal of Grit and Dust, in *Industrial Air Pollution Handbook*, ed. A. Parker, McGraw-Hill, New York, pp. 210–74.

[6] Strauss, W. (1971), *Air Pollution Control, Part I*, Wiley – Interscience, New York.

[7] Health and Safety Commission, Report of HM Alkali Inspectorate (1976), *Industrial Air Pollution*, HMSO.

[8] Turner, D. B. (1970), *Workbook of Atmospheric Dispersion Estimates*, US Public Health Service, PHS publication 999 AP26.

[9] Harrison, R. M. and Parker, J. (1977), Analysis of particulate pollutants, in *Handbook of Air Pollution Analysis* ed. R. Perry and R. J. Young, Chapman and Hall, London, pp. 84–156.

[10] Health and Safety Executive (1976), *Lead, Health Precautions*, HMSO, London.

[11] US Environmental Protection Agency (1978), *Federal Register*, **43 (194)**, 46246–77.

[12] Hesketh, H. E. (1972), *Understanding and Controlling Air Pollution*, Ann Arbor Science, Ann Arbor, Michigan, pp. 136–69.

[13] Ingwersen, J., Pederson, P. S., Nielsen, T., Larsen, E. and Fenger J. (1978), *Effects on Pollution of a Reduction or Removal of Lead Addition to Engine Fuel*, Danish National Agency of Environmental Protection, no. 131–75–10 ENV. DK.

[14] Cantwell, E. N., Jacobs, E. S. and Pierrard, J. M. (1974), Alternative Automotive Emission Control Systems, paper presented at Division of Fuel Chemistry, American Chemical Society, Los Angeles.

[15] Clayton, P., Ellis, D. J., Palmer, P. L., Potter, C. J. and Wallin, S. C. (1972), The evaluation of a filter for the removal of lead from petrol exhausts, Warren Spring Laboratory Report SR 80 (AP).

[16] US Environmental Protection Agency (1977), *Air Quality Criteria for Lead*, EPA-600/8–77–017.

[17] Health and Safety Commission (1978), *Control of Lead at Work*, Draft Regulations and Draft Approved Code of Practice, HMSO, London.

6

Control of lead discharges to water

6.1 Introduction

Lead discharges to surface waters from many industrial and domestic sources have for many years now been limited by passage of the waste waters through both industrial and sewage effluent treatment plants. To a large extent, this limitation has been fortuitous, due to the almost universal presence of lead in combination with a range of other pollutants, which have themselves required control. Thus treatment to reduce the suspended solids content of sewage in Britain typically to better than 30 mg dm^{-3}, has also brought about a substantial reduction in the lead content of the sewage (Section 6.4.2).

Similarly, in the case of industrial effluents control may be based on limitation of the discharge of another toxic metal, say copper or zinc, with a reduction in lead discharge occurring as an added bonus. In this chapter it is necessary, therefore, to consider both control methods specifically designed to limit the discharge of lead, as well as those more general methods which also affect concentrations of lead in effluents.

Before discussing the range of treatment methods available, it is necessary to examine the various standards that apply to lead in water. These set the requirement for control, if any, and subsequently provide a basis against which to monitor the effectiveness of that control.

The formulation of water quality standards is an extremely sensitive area of pollution control. It is one in which there is a tendency for political decisions to hold almost as much weight as scientific judgements. The approach to standard setting varies from one country to another and this has led to innumerable problems, particularly when international co-operation is required. This is well illustrated by the years of argument and discussion which have arisen in an attempt to reach agreement over the European Economic Communities' (EEC) Dangerous Substances Directive and the Directive on Quality of Water for Human Consumption.

Some time will therefore be spent on examining water quality standards for lead and the differing approaches to their implementation, in order to place the subsequent sections on control practices in a proper context.

6.2 Water quality criteria and standards

The terms 'water quality criteria' and 'water quality standards' are often used interchangeably. Indeed, the dictionary definition implies a similarity of meaning. However, it is useful in this field to distinguish between 'standards', as implying statutorily imposed limits, and 'criteria' which imply desirable quality characteristics related to a particular water usage and based on scientific evidence without reference to political or legislative issues [1]. On this basis 'standards' are clearly derived from 'criteria'.

Before considering specific criteria for lead it is worth noting the qualitative distinction drawn by the EEC's Directive on Dangerous Substances, which defines List 1 ('Black List') and List 2 ('Grey List') substances [2]. Priority is attached to 'Black List' substances, which include mercury and cadmium, because of their toxicity and persistence in the environment and because they tend to concentrate in living organisms. Lead is to be found on the 'Grey List', to which increasing attention will no doubt be turned in the not too distant future.

6.2.1 Lead in drinking water

The range of water quality standards and criteria applied to lead in drinking water is illustrated in Table 6.1. The concentration limits are formulated on the basis of a daily consumption of 2 dm^3 of water. This is actually somewhat above the general average consumption, thus incorporating a safety factor in the limits for the average person. However, some people are known to consume over 4 dm^3 of water daily. This clearly illustrates the problem of whether standard setting should be based on the worst possible case, or for 'Standard Person', or alternatively for the average of a critical sub-group of the population.

The World Health Organisation (WHO) are currently re-examining their standards for lead in drinking water [3]. They are basing their reassessment on the concept of 'Health Hazard'. In the case of lead they consider exposure to be best indexed by the blood lead concentration and suggest that a median population blood lead level of 20 $\mu g\ (100\ ml)^{-1}$ should be the maximum acceptable value [cf. the US EPA value of 15 $\mu g\ (100\ ml)^{-1}$ (Section 7.6.2)]. Furthermore, they identify 3 groups as especially vulnerable (a) the foetus, (b) the young infant, (c) individuals with certain medical conditions involving increased water intake or renal dialysis.

The WHO then go on to assign a high degree of research priority to a number of areas, including the nature of the physical and chemical forms of lead in drinking water, the consumer's habit of drinking water and further epidemiological studies [3]. This list highlights the degree of uncertainty to be attached to existing

Table 6.1 Drinking water standards for lead*.

	Year	Standard (mg dm^{-3})	Reference
USSR	1970	0.1	[1]
WHO			
European 1970	1970	0.1 (0.3)†	[3]
International 1971	1971	0.1	–
South Africa	1971	0.05	[1]
EEC	1975	0.05	[3]
US EPA	1975	0.05	[4]
US National Academy of Science	1977	0.025‡	[5]
Ontario, Canada	1978	0.05	[6]
Anglian Water Authority, England	1978	0.1 (0.07)§	[7]

*In some cases the limit applies to the raw water, in others to the water at the point of use, although a clear distinction is not always made.
†The higher limit is for water in contact with lead pipes for 16 h.
‡The US National Academy of Science state 'The no-observed-adverse-health-effect level cannot be set with assurance at any value greater than 0.025 mg dm^{-3}'.
§0.07 mg dm^{-3} is a maximum desirable value.

standards which may well be revised in the years to come in the light of new information.

6.2.2 Lead in freshwaters

Standards for lead in freshwaters are based upon both the protection of aquatic life and the quality of the water for public supply. The standards for public water supply have already been discussed (Section 6.2.1).

'Water Quality Objectives' have been set for lead in the North American Great Lakes in order to protect aquatic life. These are respectively 0.01, 0.02 and 0.025 mg dm^{-3} (total lead) for Lake Superior, Lake Huron and the remaining Lakes. They are based on evaluations of the highest concentration of metal causing no harmful effect to aquatic organisms [8]. They have, however, been criticized as being based on experiments with single metals. They do not therefore allow for possible additive or synergistic effects, which may require the metal to be restricted to an even lower concentration to protect sensitive aquatic life [8].

In England, the Anglian Water Authority has adopted maximum desirable and maximum permissible lead concentrations of 0.4 and 0.5 mg dm^{-3} respectively as criteria to protect rivers used for fishing (trout, dace, roach, perch, bream) [7].

The toxicity of a metal to aquatic life is closely related to the chemical form of the metal, although as yet the precise relationship between chemical form and toxicity is not clearly established. The chemical form depends in turn on other water quality parameters (Section 3.4) and is thus likely to be highly variable.

This mitigates against the simplistic application of a single standard to protect aquatic life in all waters. At the same time, it complicates the derivation, application and adminstration of standards individually designed to suit a particular water.

6.2.3 Sewage works final effluent

There is little information available on standards for lead in sewage works effluent. This may be due to low concentrations of lead normally found in treated effluent (Section 6.4.2) which are unlikely to cause the freshwater standards previously cited to be exceeded (Section 6.2.2), given a reasonable dilution of the effluent by the receiving water.

The Los Angeles District which administers the area's Joint Water Pollution Control Plant which discharges to the sea, has set an effluent requirement for lead of 0.1 mg dm^{-3} [9]. They propose to raise this, however, to 0.4 mg dm^{-3}, based on a 50:1 dilution at the ocean outfall not previously allowed for.

6.2.4 Industrial effluent

The standards applied to lead in industrial effluent depend upon whether the effluent is discharged directly to a surface water or indirectly via a sewage treatment works (Table 6.2). The use of municipal treatment plants to treat industrial effluent is favoured in Britain and elsewhere. Treatment of the mixed and diluted wastes that result can prove highly effective (Section 6.4.2). Furthermore, it is administratively easier to maintain control of polluting discharges if they are routed via one control plant.

Preliminary treatment of the industrial effluent prior to discharge to a sewer, may, nonetheless, still be necessary in order to avoid poisoning of the biological treatment processes. It is usual in these cases to apply either a limit on the concentration of an individual metal, or more commonly a limit based on the concentrations of several metals in combination.

The inhibition thresholds due to lead are reported to be 0.1 mg (Pb) dm^{-3} for the activated sludge process and 0.5 mg (Pb) dm^{-3} for the nitrification process [11]. Nitrification proceeds only to a small extent in an activated sludge plant but to a larger extent in a biological filter and the latter threshold presumably applies to this form of treatment.

The sludges that result from sewage treatment may subsequently be subjected to a period of anaerobic digestion. This can likewise be inhibited by the presence of heavy metals which show a tendency to concentrate in the sewage sludges. In order to avoid this inhibition it has been recommended that the lead concentration of the raw sewage entering a works should be limited to 7 mg dm^{-3} (or 0.067 milliequivalents dm^{-3}) or some combination of Zn, Ni, Pb, Cd and Cu not exceeding 0.067 milliequivalents dm^{-3} [21].

Table 6.2 Industrial discharge limits (total metal concentration except where indicated).

Discharge	To inland water ($mg\ dm^{-3}$)	To sewage treatment plant ($mg\ dm^{-3}$)	Reference
Multi-product dyestuff company, England	–	0.4*	[10]
Electroplating industries, US	–	0.4 (0.8)†	[11]
Discharges to Los Angeles sewer system	–	40	[9]
Discharges to sewer system, Johannesburg, SA	–	5	[12]
Discharges in Sheffield area, England	0.5	0.5	[13]
Effluent limits, Poland	0.1	0.1	[14]
Zinc–lead blast furnace, Japan	0.5	–	[15]
US EPA discharge limits	0.05	–	[16]
Swiss discharge standard	0.5	–	[17]
Hardware manufacturer, Tennessee	0.05 (0.1)‡	–	[18]
Illinois state discharge limit	0.01	–	[19]
Base metal mines			[20]
British Columbia (new mines)	0.05 (dissolved)	–	–
Yukon Territory	0.02	–	–
Ontario	<1 (Total Pb, Cu, Zn, Ni)	–	–
New Brunswick	0.1	–	–
Environmental Protection Service, Canada	0.2	–	–
US EPA	0.2	–	–

*Direct to treatment plant.

†0.4 $mg\ dm^{-3}$ limit on average daily value for 30 consecutive days, 0.8 $mg\ dm^{-3}$ maximum in any one day.

‡Average and maximum concentrations. Discharge is to a storm ditch with no dilution capacity.

6.3 Uniform emission standards versus water quality objectives

A discussion of water quality control would be incomplete without some consideration of the divergent approaches of those countries favouring 'uniform emission standards', e.g. USA, German Federal Republic, France and Italy, and those strongly opposed to this approach, e.g. the UK and South Africa, who base control on a system of water quality objectives [2, 22, 23].

6.3.1 Water quality objectives

This approach, adopted by Britain, is designed to be very flexible in operation. It is based upon the derivation of water quality objectives (concentration limits) applying to the receiving water, which are closely linked to the use of the receiving water in question, rendering it fit for all present and future uses. Local emission standards are subsequently established having regard to the particular water quality objective and taking account of the dilution capacity of the receiving water (Fig. 6.1).

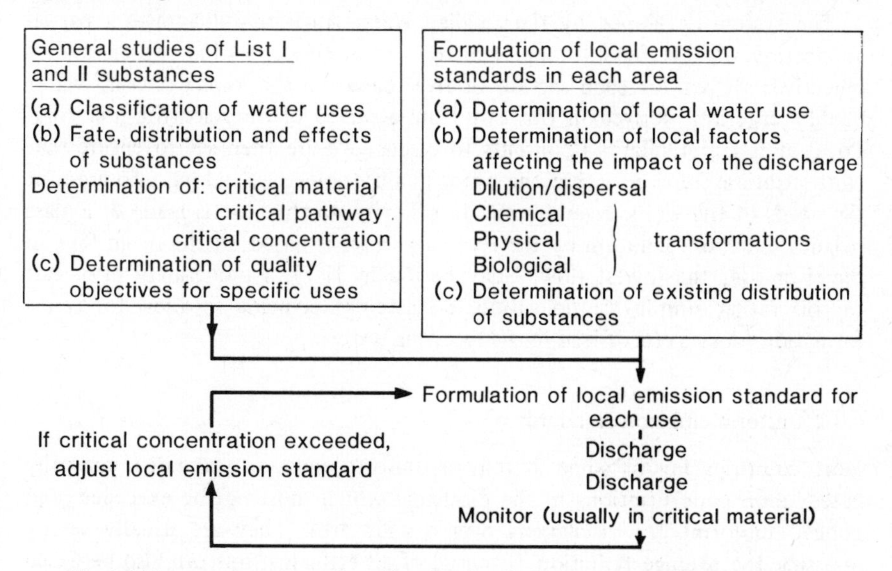

Fig. 6.1 The use of water quality objectives for discharge control: procedures to be followed, [22].

An important step in this process is the study of the fate, distribution and effects of the particular substance in the environment. Furthermore, this study should extend to all likely effects whether local to the discharge or at a distance. It should also take into account chemical, physical and biochemical transformations which might affect the substance [22]. The outcome should be to determine the most undesirable effect, known as the 'critical effect', for which there will be a 'critical material' directly related to the effect, a 'critical pathway' to that material and a 'critical concentration' in that material, be it water, fish, sediment etc.

This procedure for setting water quality objectives relies upon a considerable knowledge of the behaviour of a substance in the environment. Given the present uncertainty about many aspects of the behaviour of lead in water, as well as biological systems, this approach must obviously only be applied with qualifications and reservations. There must necessarily be many subjective judgements involved in setting a water quality objective and for this reason there is a strong case for performing the setting of water quality objectives in public [22].

Similar judgements must be made when transforming water quality objectives into local emission standards. These will be based on both the capacity of the local environment and upon the steps which it is technically and economically possible to take in order to control the discharge. This procedure has in the past been shrouded in secrecy and hence been open to considerable public criticism. However, it is now (1979) policy in Britain to publish the individual 'consents to discharge'.

The system developed by the Anglian Water Authority illustrates a partial application of this flexible approach to control (Fig. 6.2) [7]. Water quality objectives are set for each section of river based on the lowest relevant water quality criterion, derived in this case from a survey of the scientific literature. Local emission standards ('consents to discharge') are then set to ensure that concentrations remain within the water quality objectives, taking into account the needs of the whole river system. In order to do this, use is made of a mass balance equation. Dilution by the receiving water is based on the mean flow in the river over the lowest flow 7 day period in 1973, a moderately dry year. In this rather simplistic application, no attempt is made to allow for transformations, losses etc. of lead in the receiving water.

6.3.2 Uniform emission standards

Most countries favour some system of uniform emission standards, generally based upon concentrations in the discharge which must not be exceeded, and applied uniformly to dischargers over a wide area. They are usually set by assessing the average pollution potential of all effluents, but can also be based solely on the effectiveness of current treatment technology [2]. They have the advantage that they are administratively straightforward and easily understood, and because of their simplicity afford less opportunity for evasion by the discharger. A further benefit frequently claimed for uniform standards is that they impose the same uniform costs for pollution control throughout a particular industry.

In this uniform approach no account is taken of the volume, and hence mass, and location of the discharge, nor the volume and nature of the receiving water. Thus, although calculated to be correct on average, uniform standards can result in over-protection in some areas and under-protection in others [2].

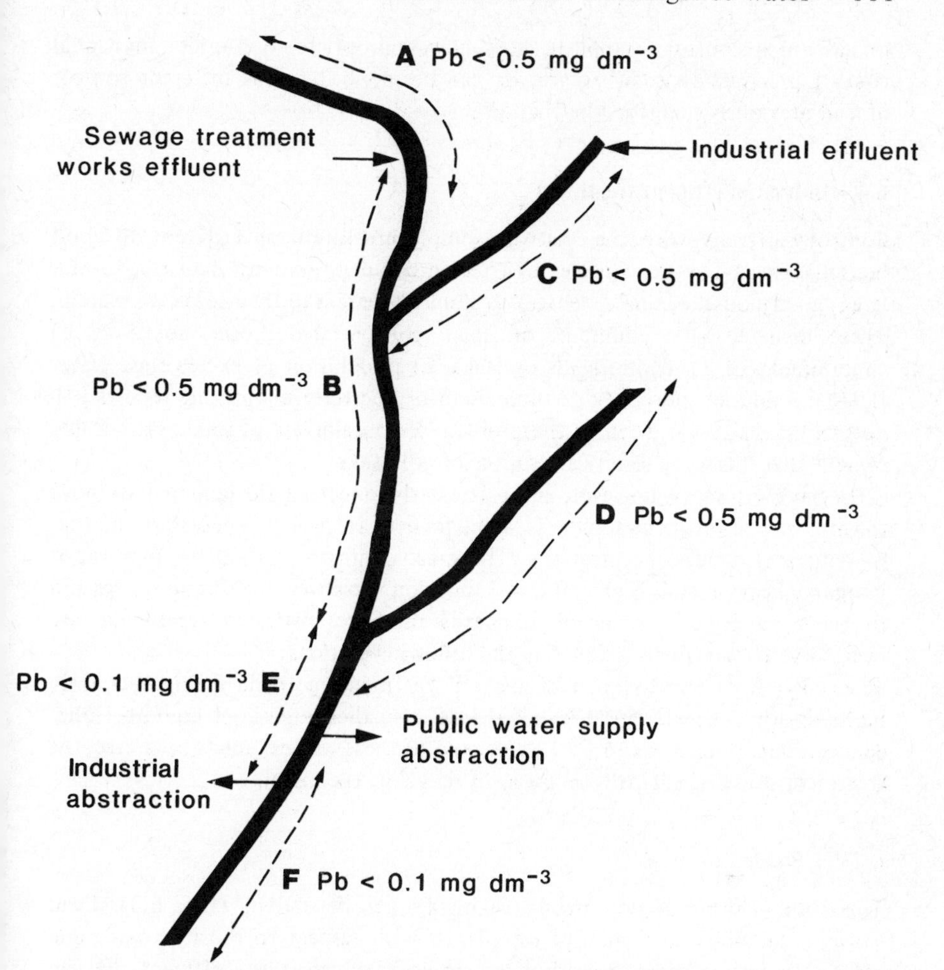

Fig. 6.2 Typical classification of river use and maximum permissible lead concentration for each stretch using Anglian Water Authority water quality criteria (based on [7]). A. Trout fishery; general amenity. B. Perch, bream fishery; general amenity; spray irrigation. C. Roach fishery; general amenity. D. Dace, perch, roach fishery; livestock watering; general amenity. E. Public water supply; perch, roach, bream fishery; general amenity. F. Industrial water supply; perch, roach, bream fishery; general amenity; livestock watering.

An example of this uniform approach is contained in the US Federal Water Pollution Control Act which is to be instituted in two stages. The initial stage is based on the application of the best practicable control technology (BPT) currently available. This progresses to the application of the best available technology (BAT) economically achievable [23]. The US EPA may also set federal standards for effluents to serve as guidelines for the above [11].

6.4 Control practices for lead discharges

Having set forth the specific requirements for control, as well as the principles under which control is applied, it is now appropriate to consider individual control practices that either are, or can be, applied to the different sources of lead previously identified in Chapter 3.

6.4.1 Industrial effluent treatment

Control can start with the relatively simple institution of a system of 'good housekeeping'. The Los Angeles District, in co-operation with industry, formulated 'good housekeeping' practices to limit discharges to their sewerage system. These included a) no dumping of spent, concentrated process solutions; b) containment of accidental spills or leaks; c) prohibition of excess rinse water flows for dilution purposes; d) submission of quarterly monitoring reports [9]. Special emphasis was given to the problem of containment of spills as it became evident that this was a significant source of pollution.

In most instances, however, and particularly in view of the tendency to move towards stricter effluent standards, some form of treatment is necessary in order to reduce effluent concentrations. The most common method for removal of inorganic heavy metals is chemical precipitation. However, for certain wastes and to meet particularly stringent standards more sophisticated treatments are necessary. These will be outlined in the following sections.

Finally, it is worth remembering all treatment methods are in fact only methods for concentrating the metal into a smaller volume of liquid or solid. The problem remains as to how to dispose of these wastes, unless of course, the treatment is used principally as a way of recycling the metal.

6.4.1.1 Precipitation

The solubility of heavy metals is highly pH dependent (Fig. 6.3). Lead shows a solubility minimum at *ca.* pH 10 with respect to the hydroxide and pH 8–10 with respect to carbonate species. Adjusting the effluent pH can therefore cause the precipitation of the metals which are subsequently removed by sedimentation or filtration. Often a flocculating or coagulating agent is added to improve the sedimentation characteristics of the precipitate. The pH chosen for the precipitation will depend upon the range of metals to be removed and must be optimized for each individual effluent. It is sometimes necessary to employ a two-stage pH treatment to remove all metals.

If space is no problem, then the sedimentation can be carried out over a matter of days in large lagoons. However, in many applications it is necessary to incorporate this in a more specialist treatment plant. A typical layout for a continuous-treatment system for heavy metals removal is shown in Fig. 6.4. Allowance is made for a portion of the sludge to be recycled to provide a seed to aid the growth of newly formed precipitate. Filtration may be advisable to ensure the maximum removal of metals. This is generally through a sand

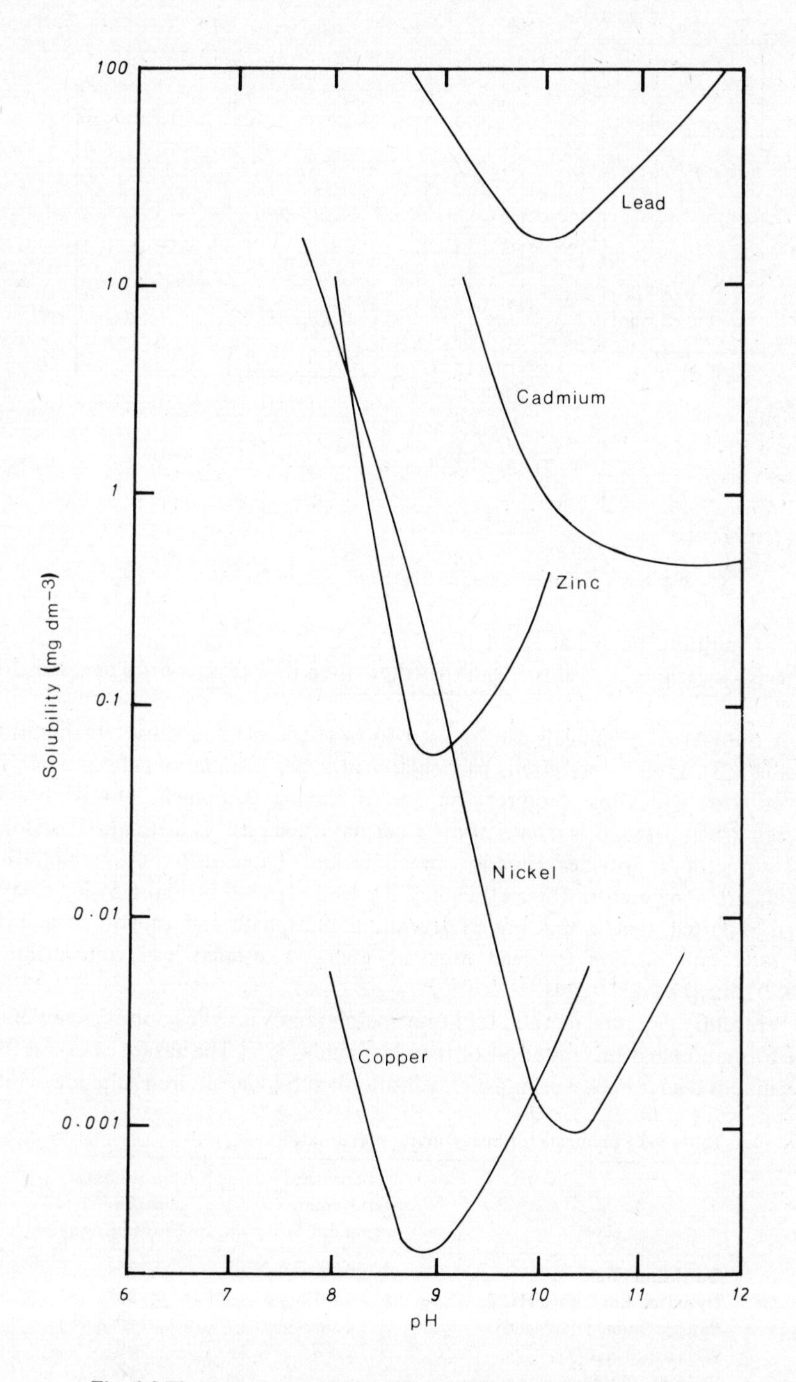

Fig. 6.3 Theoretical metal hydroxide solubility as a function of pH.

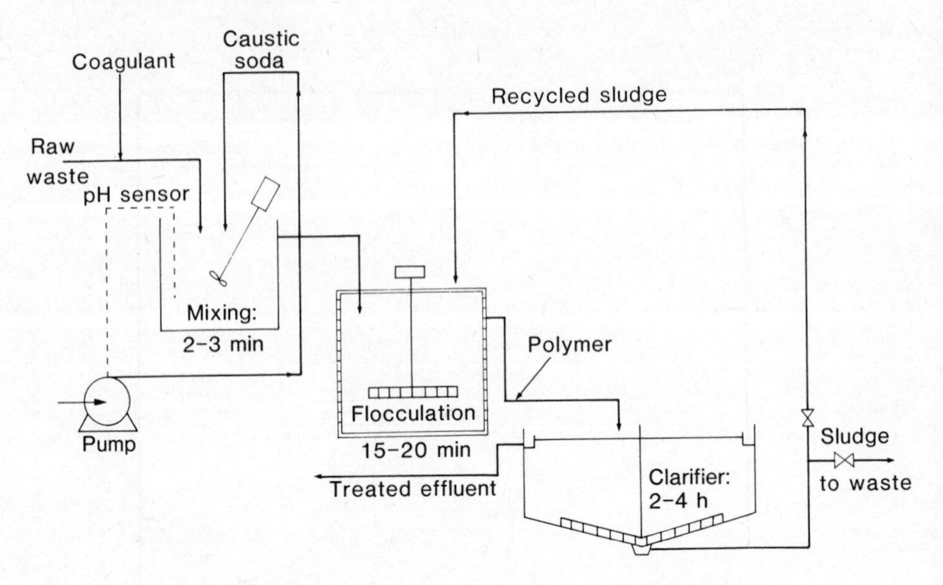

Fig. 6.4 Typical continuous treatment system for heavy metals.

bed or multi-media bed at about 10 to 30 dm^3 min^{-1} per square metre of surface area. Backwashing is used to clean the filter when the pressure drop becomes too great.

A number of chemicals can be used to raise the pH and cause precipitation (Table 6.3). Lime is preferred, particularly in larger installations, because of its lower cost. Quicklime requires the use of slaking equipment but is cheaper overall when demand is over 5 tonnes per day. Soda ash is difficult to dissolve but sometimes provides superior precipitation. Three different precipitating agents are compared in their efficiency for lead removal in Fig. 6.5. It has also been reported that a mixture of trisodium phosphate and caustic soda is an effective combination for lead removal, giving a residual lead concentration of 0.02 mg (Pb) dm^{-3} [24].

A recently patented method for heavy metals removal, called Sulfex, relies upon the formation and precipitation of metal sulphides [25]. The hazard of excess free sulphide is overcome by using a highly insoluble sulphide salt, iron sulphide, which

Table 6.3 Chemicals for heavy metal precipitation [24].

Chemical	Theoretical requirements (mg dm^{-3})	Approximate cost (1977) (\$ $tonne^{-1}$)
Quicklime, CaO	0.76	25
Hydrated lime, $Ca(OH)_2$	1.00	25
Caustic Soda, 50% NaOH	1.08	140
Soda ash, Na_2CO_3	1.42	50
Sodium sulphide	–	250

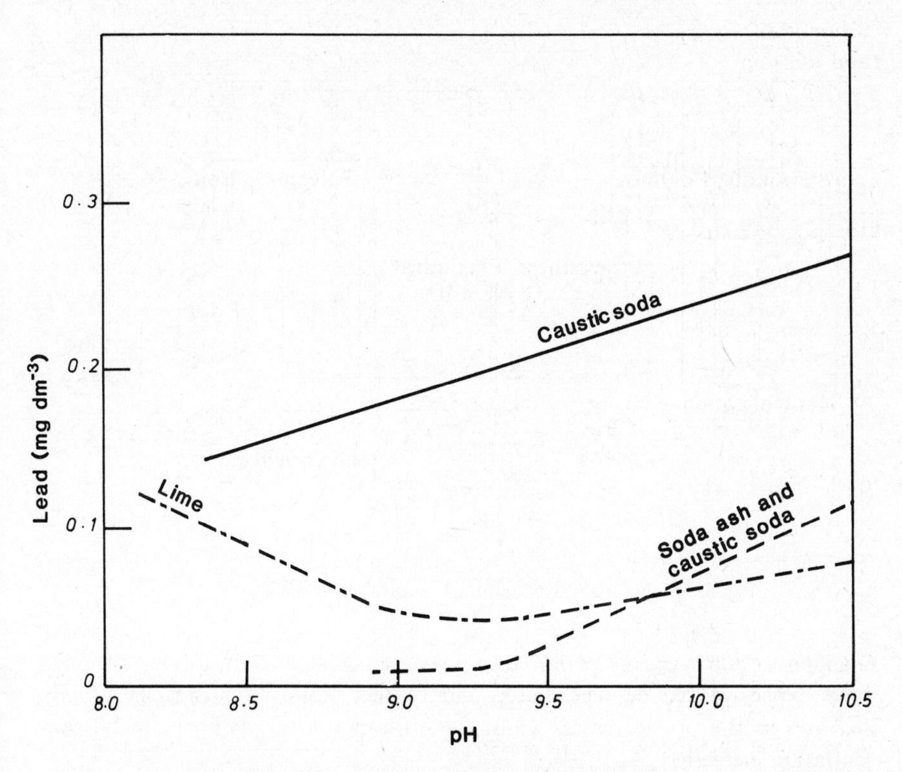

Fig. 6.5 Experimental lead solubility in three alkalies, [24].

results in a sulphide ion concentration of only 0.000 02 mg dm^{-3}. The reaction sequence then becomes in effect

$$Pb^{2+} + FeS \rightarrow PbS + Fe^{2+}$$

The theoretical solublity of lead sulphide is *ca.* 4×10^{-9} mg (Pb) dm^{-3}, far lower than that of the hydroxide or carbonate species. The process is reported to be more effective than conventional precipitation, particularly when chelating agents are present in the effluent, since these normally reduce the effectiveness of precipitation by raising the apparent solubility.

The layout of a Sulfex plant is shown in Fig. 6.6. A preliminary hydroxide precipitation reduces the amount of solids produced by the Sulfex process, as well as the amount of more expensive iron sulphide required, thus keeping down costs. Table 6.4 sets out comparative costs of treatment by hydroxide precipitation and the Sulfex process. A direct comparison is nevertheless complicated by the greater efficiency of the Sulfex process.

6.4.1.2 Activated carbon adsorption

Activated carbon will adsorb many metals which are organically complexed.

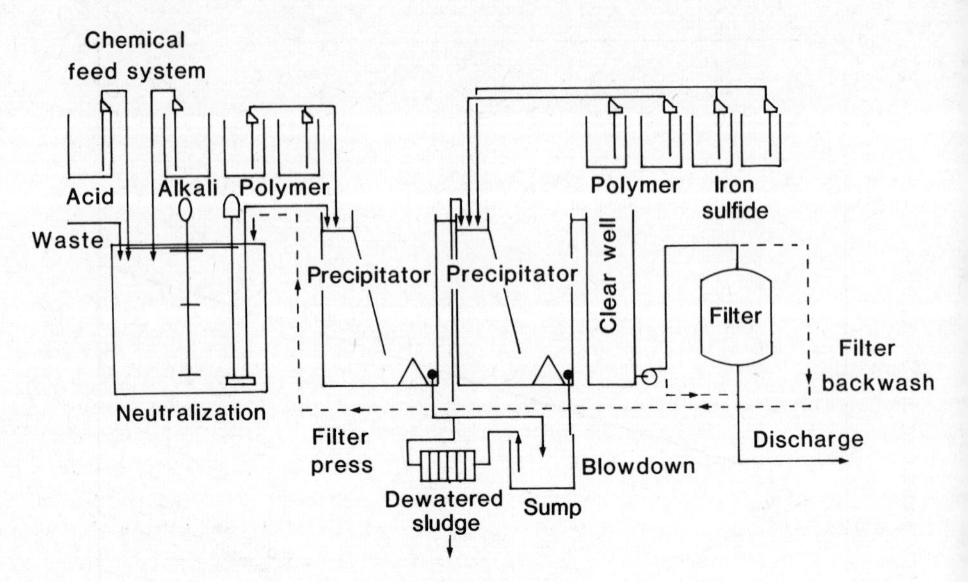

Fig. 6.6 Sulfex plant for combined removal of heavy metals, [25].

Granular carbon is usually preferred to powdered carbon, as the former, although more expensive can be regenerated and reused. Adsorption capacity usually increases as the pH decreases, whilst the efficiency increases as the metal concentration increases [24].

Table 6.4 Comparative cost of heavy metal removal by hydroxide precipitation and the Sulfex Process* (based on [25].

Process	Chemical dosage (kg $(1000\ m^3)^{-1}$)		Cost (\$ $(1000\ m^3)^{-1}$)	
Hydroxide precipitation	pH 7.5	pH 10	pH 7.5	pH 10
$Ca(OH)_2$	40	110	1.6	4.5
Polymer	4	4	4.7	6.3
H_2SO_4	–	73	–	8.2
		Total	6.3	19.0
Sulfex	pH 8.5	–	pH 8.5	–
71% NaHS	11	–	2.6	–
$FeSO_4 \cdot 7H_2O$	92	–	4.5	–
Polymer	4	–	4.7	–
$Ca(OH)_2$	136	–	5.2	–
		Total	17.0	

*Based on tests with synthetic wastes, with 4 mg dm^{-3} of Cu, Cd, Cr^{3+}, Ni and Zn at pH 6.0. The final effluent concentrations are not comparable.

6.4.1.3 Ion exchange

Soluble metals exchange rapidly with cations, such as hydrogen, sodium or calcium, saturating an ion exchange resin. Once the ion exchange resin becomes saturated by the replacement metal it can be regenerated by flushing with an acid or alkaline solution. The metals are thus concentrated into a small volume. Ion exchange is particularly applicable to waste streams which cannot be treated effectively by other means or where the metals can be recovered and reused.

6.4.1.4 Other removal methods

Reverse osmosis is a treatment system utilizing semi-permeable membranes. The system operates at pressures up to 40 atmospheres and produces a concentrate on one side of the membrane and a clear permeate on the other. It has been applied to nickel in plating-bath waters and is most effective if the concentrate can be reused [24]. Its value for treating lead-containing wastes is uncertain.

Ozone treatment can be applied to metal removal. Its effectiveness results from both the oxidation of organo-metallic complexes, which otherwise maintain a significant portion of the metal in solution, and the oxidation of the metal to a higher, and in the case of lead, less soluble oxidation state, i.e. from Pb^{2+} to Pb^{4+} [16]. Bench studies have shown that between pH 9 and 11 uncomplexed lead can be reduced in concentration to less than 0.1 mg dm^{-3} (0.45 μm filterable) in under 2 min contact time with ozone [16].

Metal removal can also be effected by cementation, a process in which dissolved metal contacts a more active metal, such as iron or zinc, and becomes cemented to it [24].

An anaerobic filter has been shown to be effective in bench-scale tests, at removing metals from a leachate with a high organic content from a landfill site [26]. In this system anaerobic bacteria attach themselves to a plastic filter medium in a column. The metals in the water passing slowly up through the column are then precipitated under the anaerobic conditions as sulphides, carbonates and hydroxides. The lead concentration in the particular leachate investigated was reduced from 0.38 mg (Pb) dm^{-3} to an average of 0.08 mg dm^{-3} (total metal) in the treated water.

Starch-based products are also being developed for heavy metals removal [19]. Starch, a natural polymer derived from agricultural crops, is both abundant and relatively inexpensive. Once converted to an insoluble starch xanthate (ISX) it can very effectively scavenge uncomplexed metals from a waste water. In a bench test ISX was dosed at 0.64 g dm^{-3} to a solution containing *ca.* 100 mg dm^{-3} of Pb^{2+} at pH 3.7. After stirring for 5 to 60 min the filterable lead concentration was down to 0.035 mg dm^{-3}, with the pH having risen to 8.9.

On a smaller scale, the 'Ferrite Process' is reported to be widely used in Japan to remove metals from laboratory wastewaters in universities and institutes [27]. The process, outlined in Fig. 6.7, relies upon the formation of magnetically

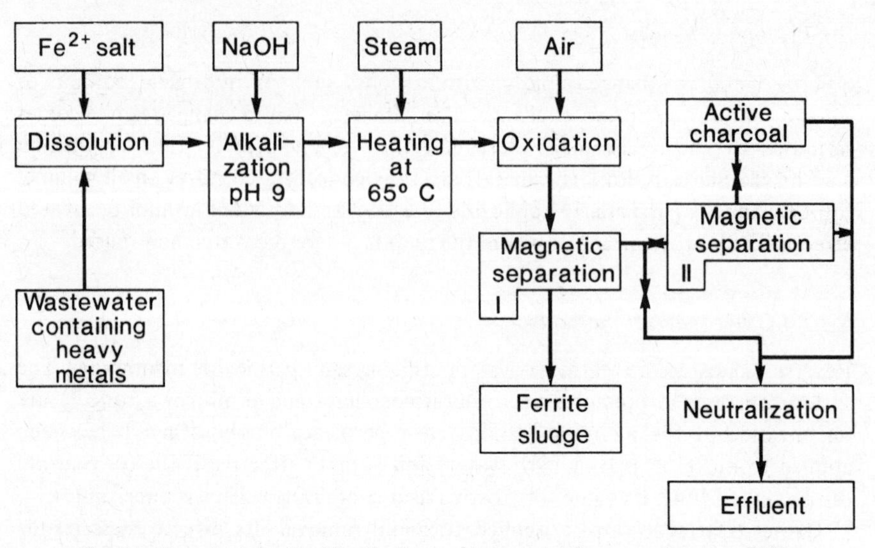

Fig. 6.7 Flowchart for the 'Ferrite Process' heavy metals removal system, [27].

susceptible Fe_3O_4 particles (0.2–0.3 μm in size). These result from the oxidation of $Fe(OH)_2$, derived from ferrous sulphate added to the wastewater, and in the process of their formation they scavenge the heavy metals. The Fe_3O_4 particles are then magnetically separated from the solution.

Other options available, but of limited significance, involve liquid–liquid extraction, electrolysis, evaporation or electrodialysis. Evaporation can result in zero discharges but is confined to dry sunny climates, where no energy input other than sunlight is required to cause evaporation.

6.4.1.5 Practical experience of lead removal

Experience has shown that theory and bench-scale tests are not always scaled up to successful effluent control. Indeed, it is reported that although bench tests give low residual lead concentrations, many operating plants are experiencing difficulty in attaining effluent lead levels below 0.5 mg dm^{-3} [24]. The following sections therefore examine practical experience of industrial effluent controls and the different solutions to the problems encountered. In many regards the discussion applies to heavy metals in general, although emphasis will be placed upon the results for lead.

(a) *Multi-product dyestuff company*

The study by Anderson and Clark [10] of heavy metals removal from the effluent of a medium-sized multi-product dyestuffs company illustrates the progressive development of an effluent control system. Typical concentrations of metals in the raw effluent and the discharge limits set by the water authority are set out in Table 6.5.

Table 6.5 Typical raw effluent and required metal discharge limits for a multi-product dyestuffs company in England [10].

	Raw effluent	Metal discharge limit
pH	1.7	–
Suspended solids (mg dm^{-3})	150	–
COD (mg dm^{-3})	2300	–
Sulphate (mg dm^{-3})	3500	–
Sulphide (mg dm^{-3})	12.5	–
Cr	18	1.0
Pb	3.0	0.4
Cu	1.5	0.4
Zn	17	0.5
Fe	19	–
Mn	26	–
Cd	–	0.005
Ni	–	0.2

The initial approach was to attempt to limit or replace the offending processes. It was, nevertheless, still necessary to employ some form of effluent control. A review of the methods available was carried out and it was concluded that precipitation would be the most suitable technique. Theoretical solubilities of the metal hydroxides were unlikely to be reliable as the effluent contained a variety of wetting agents, dispersants, phenols etc. which accounted for the high chemical oxygen demand (COD) values. Thus bench-scale tests were undertaken to establish the optimum chemical dosing rate and pH. The effect of 3 precipitation agents was examined (Table 6.6). In these tests sodium oleate was added at 3 ml dm^{-3} of a 10% w/v solution to act as a collector (coagulant) and the precipitated solids removed by flotation.

These preliminary tests provided the basis for a pilot-plant investigation. The variable nature of the plant effluent required a stagewise pH adjustment from

Table 6.6 Bench-scale test for heavy metal removal (based on [10]).

Precipitating agent	pH	Sludge weight (g dm^{-3})	Residual metal concentration (mg dm^{-3})			
			Cr	Pb	Cu	Zn
25% Soda ash	7	16	<0.2	0.3	<0.05	0.51
	8	19	0.7	0.3	<0.05	<0.05
	9	17	<0.2	0.3	<0.05	<0.05
10% Lime slurry	7	9	0.9	0.3	<0.05	0.56
	8	7	<0.2	0.3	<0.05	<0.05
	9	13	<0.2	0.3	<0.05	<0.05
Solid lime	7	7	<0.2	NT	<0.05	<0.05
	8	16	<0.2	NT	<0.05	<0.05
	9	15	<0.2	NT	<0.05	0.1

NT = No trace

pH 1.5 to pH 9, using a 10% w/v lime slurry. In the pilot-plant the precipitate was removed by settlement. Efficient settling could generally be achieved with a retention time of 4 to 6 h, without the addition of polyelectrolyte flocculants. Sludge recycle was incorporated to improve the precipitation rate, with the result that low effluent metal concentrations were achieved (Table 6.7).

Table 6.7 Pilot-plant removal of heavy metals using 10% lime slurry and sludge recycle (based on [10]).

Influent				Effluent					
						Metals ($mg\ dm^{-3}$)			
Experiment	pH	Suspended solids ($mg\ dm^{-3}$)	COD ($mg\ dm^{-3}$)	pH	Suspended solids ($mg\ dm^{-3}$)	Cr	Pb	Cu	Zn
1	1.5	5500	3100	9.3	30	0.3	0.3	0.1	0.6
2	1.2	6500	2700	9.4	20	0.2	0.4	0.2	0.4
3	1.6	5850	1900	9.5	25	0.25	0.2	0.15	0.3

The next step was to scale up from the pilot-plant, which could handle $0.9\ m^3h^{-3}$ of acid effluent to a full-scale plant to handle $286\ m^3h^{-3}$ of acid effluent. The variable effluent flow required a flow balancing system. This was achieved within the neutralization unit by using a four-stage stirred reactor with a retention time of 2.5 h. Economics dictated the use of quick-lime, thus a lime storage and slaking plant were incorporated in the final design to produce $900\ kg\,h^{-1}$ of 10% w/v lime slurry. The total treatment plant cost was of the order of £700 000 (1978 price, *ca.* $1.5m).

(b) *Base metal mines*

Heavy metal-laden effluents arise from mine drainage water, surface drainage and mill process waters. The most common treatment method is to discharge the wastes to a tailings pond in which the pH is controlled within the range 9.5 to 10.5. A minimum retention of 5 days is recommended with a pond of 4 to 11 ha for each 1000 tonnes of solids discharged per day [20]. A well controlled tailings pond can achieve effluent lead concentrations of $0.1\ mg\ dm^{-3}$.

(c) *Hardware manufacturer*

Additional treatment proved necessary in order to meet the stringent requirements imposed by the State of Tennessee on a hardware company discharging metals from plating rinse waters [18]. Initial bench-scale studies were designed to optimize the hydroxide precipitation process. Addition of a high molecular weight anionic polymer proved an effective flocculating agent, but residual concentrations of some metals were still too high. The lead concentrations were, nevertheless, reduced from raw effluent levels of $0.2\text{–}2.0\ mg\ dm^{-3}$ to $<0.05\ mg\ dm^{-3}$ by precipitation alone. A cation exchange resin was then used to reduce the concentrations of the other metals still further. This was found to be most effective when used as a two-stage system, with the effluent passing first through a column with the resin in a hydrogen form, then a column in the sodium form. Unfortunately no residual lead concentrations are reported following this additional treatment.

(d) *Zinc–lead smelter*
Smith and co-workers [28] report on the effluent treatment for a zinc–lead smelter in the New South Wales, Australia. Treatment was by precipitation using a multi-stage addition of 15% w/v lime slurry to give pH 8.9 resulting in the typical final effluent concentrations shown in Table 6.8. It was necessary, however, to upgrade the system in order to meet stricter requirements.

Table 6.8 Typical analysis of a lead–zinc smelter final effluent [28].

	Concentration ($mg\ dm^{-3}$)	Discharge ($kg\ day^{-1}$)
pH	7.7	–
Lead	0.23	11
Zinc	0.86	43
Cadmium	0.17	8.5

Upgrading of the treatment was based upon both previous experience and bench-scale tests. The new plant would effect precipitation in a six tank cascade system, using a 4% lime slurry to raise the pH from 4 to 8.5–8.7 (Fig. 6.8). Flocculants and conditioners would then be added and the effluent passed to a thickener tank. From here the effluent would pass at up to 120 m^3h^{-1} to a retention tank where the pH would be raised to pH 11–12, before final settling in a reactivator. Detailed tests showed that the precipitated lead was colloidal in character and hence difficult to settle. Addition and subsequent precipitation of iron improved the settling rate, but iron addition was not planned for the upgraded treatment plant.

As the authors note, scale up from bench-scale tests to full-scale operation can introduce difficulties and a degree of flexibility must be incorporated in any design.

(e) *Electroplating wastes*
Lead is plated onto metals such as steel, aluminium and copper to improve their solderability, coating properties and performance [29]. The plating bath usually contains fluoroboric acid and boric acid, whilst materials such as glue, resorcinol, gelatin and sometimes hydroquinone are added to improve deposition.

The entire electroplating process generates complex wastewater streams. For the purpose of treatment these can be considered to occur in four segments – acid, alkali, chromium and cyanide streams [29]. The acid and alkali wastes are combined and the pH adjusted. The chromium is reduced from Cr^{6+} to Cr^{3+} with sulphur dioxide at pH 2–3. The cyanides are destroyed with chlorine or hypochlorite. Destruction of cyanides is essential for effective precipitation of the heavy metals, as they form strong complexes with the metals and thereby increase

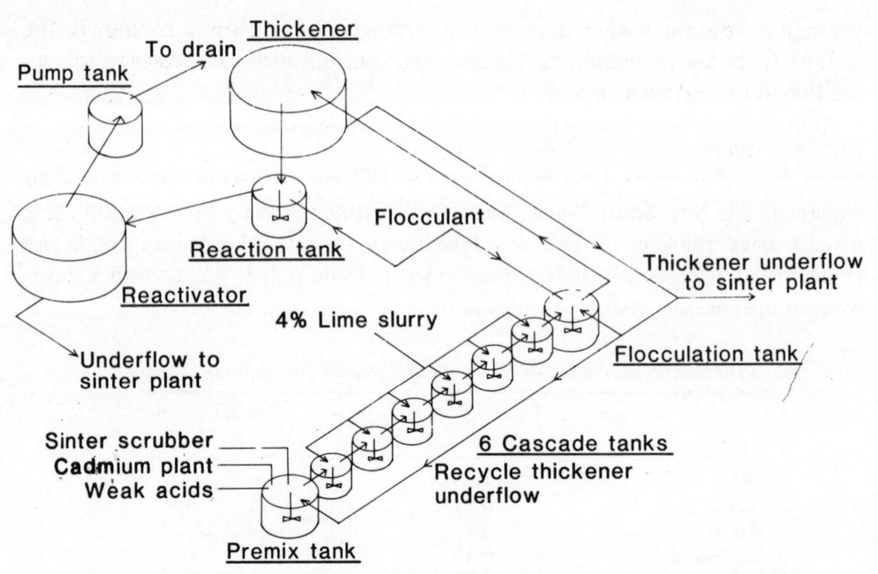

Fig. 6.8 Precipitation treatment of wastewater from a zinc–lead smelter, [28].

the effective solubility [30]. A simplified process flow diagram is illustrated in Fig. 6.9.

(f) *Tetraalkyllead manufacture*

The waste streams from a tetraalkyllead manufacturing plant contain significant concentrations of both inorganic and organic lead. The inorganic lead can readily be treated by conventional precipitation but this has little or no effect on the organic lead, which is usually present as soluble alkyllead salts. A number of methods have been proposed to reduce the organic lead concentrations, typically 1–100 mg (Pb) dm^{-3}, to less than 1 mg dm^{-3}.

A system proposed by Japanese workers [31] treats the inorganic and organic lead streams separately (Fig. 6.10). The organic lead, at up to 200 mg dm^{-3}, is removed on an Amberlite 200 ion exchange resin in the hydrogen form. The resin is eluted with sodium hydroxide and the organic lead converted to inorganic lead by chlorination.

The Ethyl Corporation [32] have proposed the use of sodium borohydride, a strong reducing agent, in the presence of Fe^{2+} ions supplied by $FeSO_4 \cdot 7H_2O$. The preferred pH range is 9–11 and addition of a water-soluble high molecular weight anionic polyelectrolyte aids the precipitation of the insoluble inorganic lead formed.

The Associated Octel Company have on the other hand patented a method involving cementation [33]. The effluent at pH 7.5–9 is passed down a column containing clean particles of zinc. Extraction efficiencies of 80–90% are maintained over long periods, 120–160 days. Occasional backwashing, sufficient to fluidize the zinc, is necessary to maintain the effectiveness of the zinc. Clean zinc particles are essential to the process and can be obtained by pre-washing with a

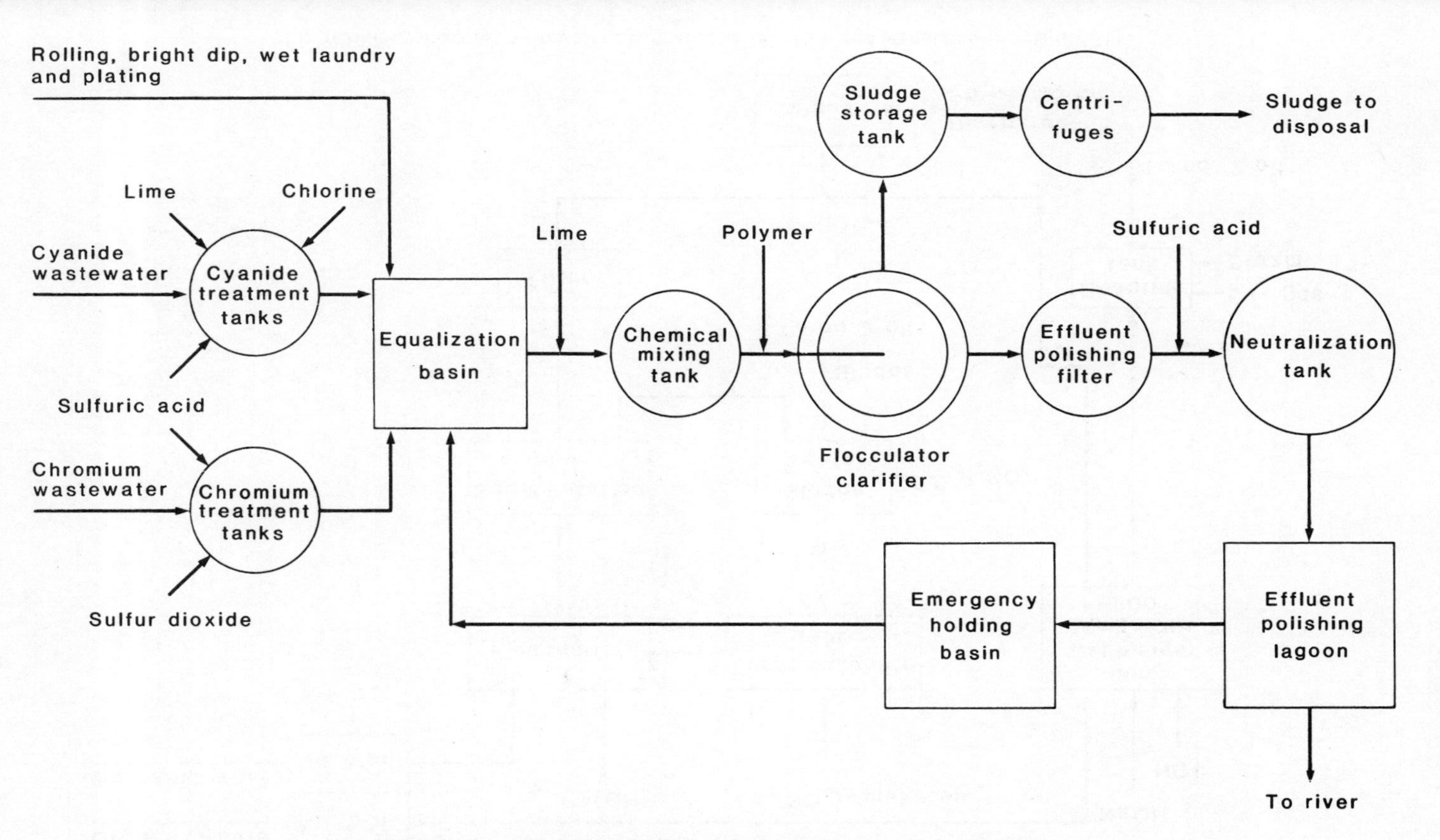

Fig. 6.9 Flowchart for the treatment of electroplating wastewaters, [30].

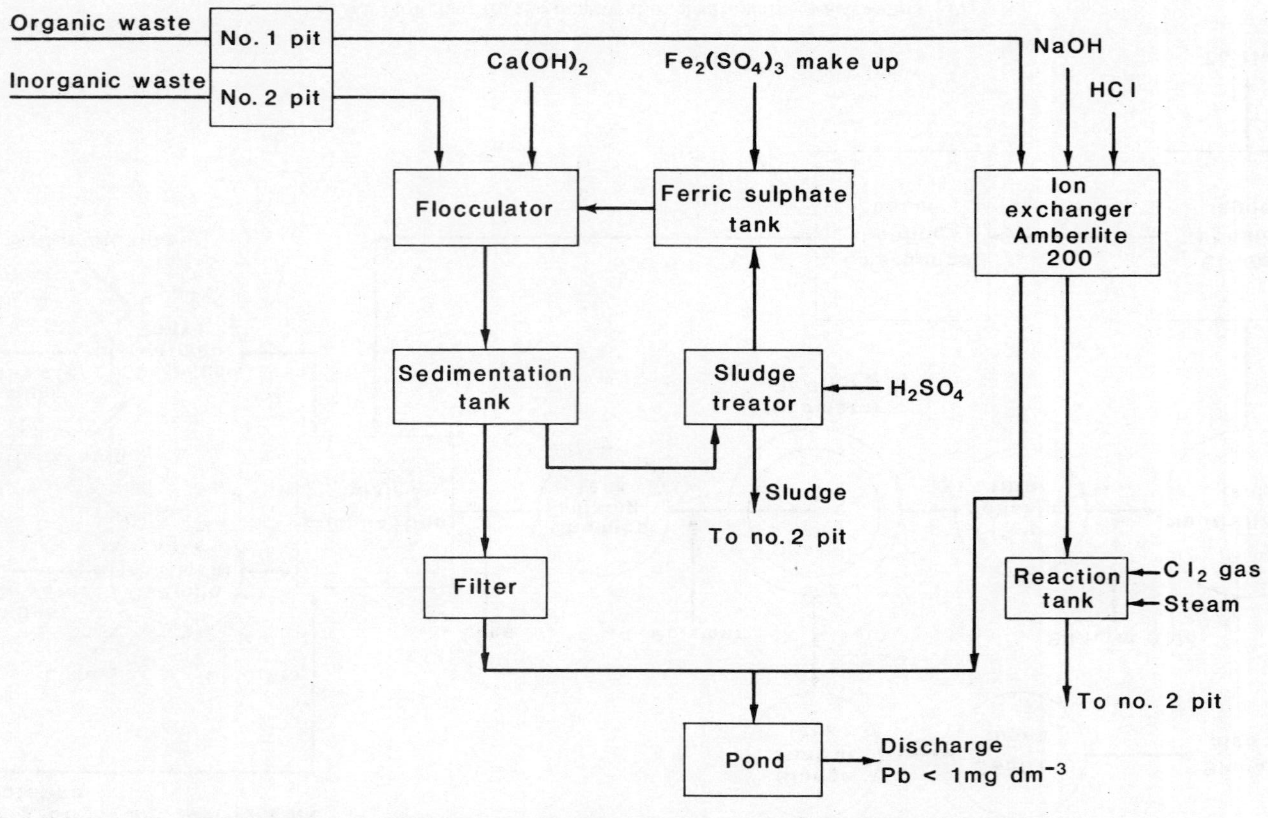

Fig. 6.10 Flowchart for wastewater treatment at a tetraalkyllead manufacturing plant, [31].

dilute acid. Tests on an actual effluent, with 10-15 mg dm^{-3} of organic lead, produced a treated effluent usually with $<$2 mg (Pb) dm^{-3} of lead.

A further recently patented method [34] employs activated carbon, which has been shown to reduce organic lead concentrations from 10-20 mg (Pb) dm^{-3} to 0.1-0.8 mg (Pb) dm^{-3}. Contact time with the activated carbon is 12-30 min and the carbon is steadily removed from the carbon bed as it becomes used up and is replaced with fresh activated carbon.

6.4.2 Sewage treatment

Sewage treatment plants are very efficient at removing lead and other heavy metals from raw sewage. They thus act as effective controls on lead discharges not only within domestic sewage but also from industrial sources and urban stormwater runoff, both of which are frequently directed to sewage treatment plants.

The layout of a typical sewage treatment works is shown in Fig. 6.11, with treatment in this case based on the activated sludge process. Metals removal occurs during both the primary sedimentation and biological treatment stages [35]. Primary sedimentation is effective for lead as a high proportion of the lead is found in the particulate form in the raw sewage. Removal efficiencies for lead will depend upon the design and maintenance of the sedimentation tanks, but generally lie around 70%.

Biological treatment by either the activated sludge process or the trickling filter causes lead and other metals to precipitate, as well as to become adsorbed and incorporated in the biomass. The lead is thus removed with the sewage sludge. Sewage sludges are often subjected to anaerobic digestion prior to dewatering. The attempted speciation work of Stover and co-workers [36] suggests that during this stage the lead becomes associated in the following order: carbonates $>$ organic bound $>$ sulphides $>$ adsorbed $>$ ion exchangeable. Removal efficiences during the activated sludge process and subsequent clarification are generally round 70%. A similar removal efficiency for lead is reported for a biological (trickling) filter system [37].

The overall efficiency for lead removal during passage through a sewage treatment plant incorporating primary sedimentation and biological treatment is usually of the order of 85-95%, whilst final effluent concentrations are generally in the range 0.001-0.030 mg dm^{-3} (Section 3.3.1).

Sewage treatment plants thus prove an effective means of reducing lead discharges to surface waters. Care must be taken, however, to avoid poisoning of the biological treatment by limiting the heavy metal input to the plant (Section 6.2.4).

Finally it is worth noting that using sewage treatment plants to control lead discharges, especially from industrial sources (Section 6.2.4), will result in elevated metal concentrations in the sewage sludge. This sludge is a useful fertilizer if returned to the land, but its value is reduced if it is highly contaminated with metals (Section 4.2.4).

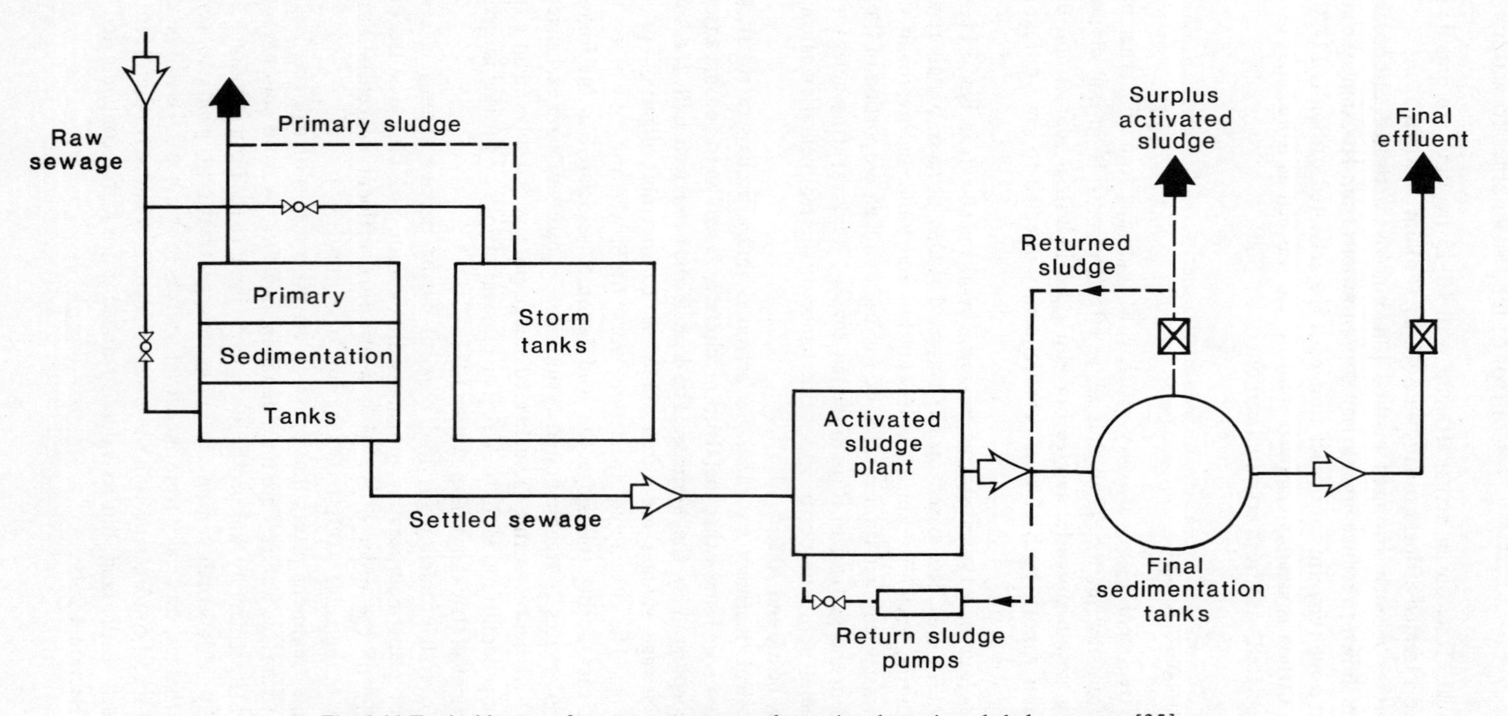

Fig. 6.11 Typical layout of a sewage treatment plant using the activated sludge process [35].

6.4.3 Stormwater control

Stormwater runoff from urban areas and highways is an important source of lead (Section 3.3.1). In many cases there has been little or no attempt to control these discharges, although the situation is changing with the growing recognition that urban and highway runoff is a major source of pollution, not only for lead, but for a whole range of substances [38].

Identification of contaminant sources and an understanding of their build-up on paved areas are pre-requisites to effective control [38]. In the case of lead the major source is the burning of leaded petrol (Section 3.2.2). Thus regulation of lead in petrol (Section 5.4.1) will result in a reduction of lead discharges. Control of pollutants in stormwater runoff will, nevertheless, have to be effected, regardless of whether lead is removed from petrol. It is thus still worthwhile examining the range of control practices available and looking at their effectiveness in limiting lead discharges.

Stormwater runoff has three important characteristics which must be taken into account when considering control measures. 1) It is an intermittent phenomenon, which produces shock loadings on any treatment facility if unregulated; 2) It is often characterized by a first flush effect, during which extremely high concentrations occur; 3) For historical reasons, it is sometimes routed by way of combined sewers to sewage treatment plants. The sewerage system often has an overflow facility to accommodate excessive flows, which are then usually discharged untreated directly to a river.

Control can be usefully considered in two categories a) non-structural or low-structural source controls – these can be called 'best management practices' (BMPs) and b) structural controls, in effect end-of-pipe treatment.

6.4.3.1 Best management practices

(a) *Street sweeping*

A large part of the lead deposited in the street surface from vehicle exhausts becomes incorporated in the street dust. It is then transported during a runoff event as a suspended solid. Street sweeping, often vacuum assisted, removes solids from the street surface and thus reduces the lead available for runoff. There remains, nevertheless, the problem of disposal of the contaminated street dust.

The US EPA have examined the effectiveness of street sweeping programmes [39]. Performance depends on the character of the street surface, the initial loading and other environmental factors. It also depends upon the number of passes of the sweeper and the interval between cleaning. Removal of the street dust in this particular study was 20–50% of the initial loading, with a more efficient removal of the larger particle sizes. Lead shows a tendency, however, to be associated with the smaller particle sizes, thus reducing the effectiveness of street sweeping for lead removal.

The cost of street cleaning in San Jose in the US in 1977 was \$10 per km swept, with labour costs accounting for 65% of the total. The average removal of lead was 0.23 kg per km swept at a cost of \$12.5 per kg removed [39]. Pitt [39] argues that street sweeping could be more cost effective than the installation of special treatment facilities, although further study is required. It is worth pointing out, however, that street sweeping is not practical on all streets, particularly where there is on-street parking and traffic congestion [40].

(b) *Alternative BMPs*
The BMP approach to control is well illustrated in the study made in order to improve the management of the combined sewer system in the City of Filchburg, Massachusetts [40]. The requirement in this case was to remove the pollution problem due to combined sewer overflow into a river during a storm event.

A range of remedial actions were considered, including a) sewer flushing to remove material deposited in the sewer during dry weather flow; b) street sweeping; c) inflow reduction from illegal downspout connections to the sewer; d) off-line storage either as small upstream modules or a large downstream facility; e) minor structural and piping modifications to the sewerage system. The latter involved raising stormwater overflow weirs and separating off two stormwater-only sewers for discharge direct to the river. The cost effectiveness of the different management alternatives is shown in Table 6.9. In this case, the best plan, as judged by Biological Oxygen Demand (BOD) reduction, is seen to be minor modifications to the overflow structures and storm sewer piping. The result is an increased flow at the sewage treatment plant of 3.1×10^4 m^3 per year, which is acceptable in this particular case.

Other BMPs involve proper maintenance of catchbasins and gully pots used to trap sediments and oils, as well as land use planning opportunities, such as avoidance of steep slopes for development, utilization of porous pavements, minimizing the impervious area etc. [41].

6.4.3.2 Structural controls

These involve storage and treatment of stormwater runoff. Storage is frequently used to attenuate peak flows, so as not to overload the treatment facility. However, it can also be considered as a treatment process in its own right, due to the sedimentation that occurs during storage.

Physical, as against biological, treatment is more effective at handling the highly variable flow rates and concentrations of stormwaters. The costs and effectiveness of different physical treatment systems for removal of suspended solids are illustrated in Table 6.10. The latter will approximate the lead removal efficiencies, for which no data are available. The swirl concentrator is clearly the most cost effective and it is currently undergoing development in the US. In this system a circular movement imparted on the inflow water concentrates the solids in a small volume of water which is separated off and can be directed to

Table 6.9 Alternative combined sewer management plans [39].

	Plan			
	A	B	C	D
Combined sewer overflow reduction				
(m^3 y^{-1}) × 10^{-5}	1.9	1.9	2.2	2.6
(%)	41	41	47	56
Reduction of BOD loading in sewer overflow (%)	44	54	58	67
Present value costs (20y, 8%) (million $)	0.026	2.2	2.5	4.7

Plan A: Minor modifications to stormwater overflow structures and storm sewer piping.
Plan B: Plan A + sewer flushing + street sweeping.
Plan C: Plan B + inflow correction.
Plan D: Plan C + 1200 m^3 storage.

Table 6.10 Comparison of physical treatment systems (based on [42]).

Treatment	Suspended solids removal (%)	Average cost ($ m^{-3} day^{-1})
Sedimentation		
Without chemicals	20–60	5.0
Chemically assisted	68	5.0
Swirl concentrator	40–60	1.0
Screening	10–90	4.2
Dissolved air flotation*	45–85	7.5
High rate filtration†	50–80	10.5

*Including prescreening and chemical addition.
†Including chemical addition.

a sewage treatment works. Clean water flows over a circular weir at the centre of the concentrator. The system has no moving parts, operates efficiently over a range of flow rates and requires only a fraction of the detention time necessary for sedimentation.

It is necessary, nevertheless, to set the cost and effectiveness of a system specifically designed to treat stormwater runoff against the alternative of merely storing the stormwater, prior to treatment in a sewage treatment plant during the subsequent dry weather period.

Finally, it should be recognized that the approach to stormwater control must be individually designed to suit each particular geographic area. There is no best solution to the problem.

6.4.4 Control of lead in drinking water

Lead occurs in raw water abstracted for the public supply system and may also become incorporated at some stage during distribution to the consumer. As previously noted (Section 3.3.5) concentrations in the raw water are generally low. They can, nevertheless, be reduced further during conventional water treatment (Table 6.11).

Table 6.11 Average percentage removal of lead by water treatment plants (based on [41]).

Process	Average removal (%)*
Microstrainers	3
Clarifiers	27
Filters	29
Treatment plants	32

*Results from 12 treatment plants. Most plants showed a minimal removal of lead. Average raw water lead concentration 0.012 mg dm^{-3}.

The major source of high concentrations of lead in drinking water is the dissolution of lead from lead pipes or lead solders used in the distribution system, known as plumbosolvency (Section 3.3.5). In Britain, it is estimated that roughly 7 million households have lead somewhere between the water main and the tap. Many households may thus be exposed to elevated levels of lead in their drinking water (Section 3.3.5). Attention is currently focused on possible remedial actions.

The only way to guarantee low lead levels in drinking water is to replace all lead pipes. However, the most recent estimate (1979) for Britain is that this would cost around £2500 million (*ca.* $5000 million) and take many years to complete [43]. A cheaper alternative would be to treat the water to reduce plumbosolvency, essentially by producing a relatively inert coating to the lead pipe. This could be achieved by causing a precipitate of basic lead carbonate ($Pb(OH)_2 \cdot 2PbCO_3$) to line the pipe, brought about by raising the pH and alkalinity of the water. This approach, however, cannot guarantee low lead concentrations especially in waters which already have an alkalinity greater than 100 mg dm^{-3} expressed as ($CaCO_3$) and as yet there is no wholly satisfactory treatment approach to the problem [43].

Pipe lining is being considered but is not a favoured option at present, as there are no proven techniques. Treatment at the tap also has many practical disadvantages, as has flushing of water before use, although the latter is a useful interim measure.

There is clearly no easy solution to the problem of lead in drinking water. Treatment may eventually prove to be the best option but a better understanding of the chemistry of lead in water will be necessary before success can be reasonably guaranteed.

References

[1] Wells, R. J. (1978), Water Quality Criteria and Standards, *Water Pollut. Control,* 77, 25–30.

[2] Water Research Centre (1976), Emission Standards in Relation to Water Quality Objectives, *Water Res. Centre Tech. Rep.*, TR17.

[3] World Health Organisation (1978), *Health Hazards from Drinking-Water*, Report on a Working Group, London, Sept. 1977, WHO, Copenhagen.

[4] US Environmental Protection Agency (1975), National Interim Primary Drinking Water Regulations, *Federal Register,* 40, 59566–88.

[5] National Academy of Sciences (1977), *Summary Report: Drinking Water and Health*, National Academy of Sciences, Washington DC.

[6] Ministry of the Environment (1978), *Drinking Water Objectives*, Information Services Branch, Toronto.

[7] Price, D. R. H. and Pearson, M. J. (1979), The Derivation of Quality Conditions for Effluents Discharged to Freshwaters, *Water Pollut. Control,* 78, 118–38.

[8] Wong, P. T. S., Chau, Y. K. and Luxon, P. L. (1978), Toxicity of a Mixture of Metals on Freshwater Algae, *J. Fish. Res. Board Can.*, 35, 479–81.

[9] Eason, J. E., Kremer, J. G. and Dryden, F. D. (1978), Industrial Waste Control in Los Angeles County, *J. Water Pollut. Control Fed.*, 50, 672–77.

[10] Anderson, D. and Clark, R. (1979), Development of a Pretreatment Process for Toxic Metal Removal, *Proceedings of the Effluent Water Treatment Convention*, Nov. 1978, Birmingham, UK, Brintex Exhibitions Ltd., London.

[11] US Environmental Protection Agency (1978), Electroplating Point Source Category, Proposed Pretreatment Standards for Existing Sources, *Federal Register,* 43, 6560–73.

[12] Funke, J. W. (1975), Metals in Urban Drainage Systems and Their Effect on the Potential Reuse of Purified Sewage, *Water SA,* 1, 36–44.

[13] Tench, H. B. (1977), Effluent Standards with Special Reference to the River Don System, *Prog. Water Tech.*, 8, 41–7.

[14] Kieszkowski, M. and Jackson, G. S. (1978), A Clean Water Project in Poland, *Environ. Sci. Technol.*, 12, 896–9.

[15] Kinohita, H., Higashitsuji, Y. and Temple, D. (1976), Use of the Zinc Lead Blast Furnace in Japan, *Proceedings of the Joint MMIJ-AIME Meeting*, Sept. 1976, Denver, Colorado, Vol. 2, The Mining and Metallurgical Institute of Japan, Tokyo and American Institute of Mining, Metallurgical and Petroleum Engineers, New York, pp. 630–46.

[16] Shambaugh, R. L. and Melnyk, P. B. (1978), Removal of Heavy Metals Via Ozonation, *J. Water Pollut. Control Fed.*, 50, 113–21.

[17] Roberts, P. V., Dauber, L., Novak, B. and Zobrist, J. (1977), Pollutant Loadings in Urban Storm Water, *Prog. Water Technol.*, 8, 93–101.

[18] Yeats, A. R. (1978), Ion Exchange Selectively Removes Heavy Metals from Mixed Plating Wastes, *Proceedings of the 32nd Industrial Waste Conference*, May 1977, Purdue University, Ann Arbor Science, Ann Arbor, Michigan, pp. 467–76.

[19] Wing, R. E. and Rayford, W. E. (1978), Heavy Metal Removal Processes for Plating Rinse Waters, *Proceedings of the 32nd Industrial Waste Conference*, May, 1977, Purdue University, Ann Arbor Science, Ann Arbor, Michigan, pp. 838–52.

[20] Bell, A. V. (1976), Waste Controls at Base Metal Mines, *Environ. Sci. Technol.*, 10, 130–5.

[21] Mosey, F. E. (1976), Assessment of the Maximum Concentration of Heavy Metals in Crude Sewage which will not Inhibit the Anaerobic Digestion of Sludge, *Water Pollut. Control,* 75, 10–20.

[22] Otter, R. J. (1979), EEC 'Dangerous Substances' Directive, *Chem. Indstr.*, May, 302–5.

[23] Cusell, F. (1979), Fixed Emission Standards With Reference to the EEC Directive, *Chem. Indstr.*, May, 306–8.

[24] Lanouette, K. H. (1977), Heavy Metals Removal, *Chem. Eng.*, Oct., 73–80.
[25] Scott, M. C. (1978), Sulfex® – A new Process Technology for Removal of Heavy Metals from Waste Streams, *Proceedings of the 32nd Industrial Waste Conference*, May, 1977, Purdue University, Ann Arbor Science, Ann Arbor, Michigan, pp. 622–29.
[26] Chian, E. S. K. and De Walle, F. B. (1978), Removal of Heavy Metals from a Fatty Acid Wastewater With a Completely Mixed Anaerobic Filter, *Proceedings of the 32nd Industrial Waste Conference,* May, 1977, Purdue University, Ann Arbor Science, Ann Arbor, Michigan, pp. 920–28.
[27] Tamaura, Y., Chyo, G. S. and Katsura, T. (1979), The Fe_3O_4 Formation by the 'Ferrite Process': Oxidation of the Reactive $Fe(OH)_2$ Suspension Induced by Sucrose, *Water Res.,* **13**, 21–31.
[28] Smith, B. C., Coppock, B. W., Scott, T. C. and Firkin, G. R. (1975), Pollution Control on an ISP Complex, *Proceedings of the South Australian Conference, Adelaide, Port Pirie, Part A*, Australasian Institute of Mining and Metallurgy, Victoria, Australia, pp. 277–87.
[29] Jakobsen, K. and Laska, R. (1977), Advanced Treatment Methods for Electroplating Wastes, *Pollut. Eng.,* 9, 42–6.
[30] Weston, R. F. and Morrell, R. A. (1977), Treatment of Water and Wastewater for Removal of Heavy Metals, in *Viruses and Trace Contaminants in Water and Waste Water* (ed. J. A. Borchardt, J. K. Cleland, W. J. Redman, and G. Oliver), Ann Arbor Science, Ann Arbor, Michigan, pp. 167–82.
[31] Nozaki, M. and Hatotani, H. (1967), Treatment of Tetraethyl Lead Manufacturing Wastes, *Water Res.,* **1**, 167–77.
[32] Ethyl Corporation (1975), Reduction of Dissolved Organic Lead Content in Aqueous Solutions, *Patent Specification 1405080*, Sept., 1975.
[33] The Associated Octel Company Limited (1975), Treatment of Dilute Solutions of Organo-Lead Ions to Reduce the Lead Content Thereof, *Patent Specification 1417078*, Dec., 1975.
[34] Otto, J. M. (1978), Method of Removing Dissolved Organo-Lead Compounds from Water, *US Patent 4070282*, Jan., 1978.
[35] Lester, J. N., Harrison, R. M. and Perry, R. (1979), The Balance of Heavy Metals Through a Sewage Treatment Works. 1. Lead, Cadmium and Copper, *Sci. Tot. Environ.,* **12**, 13–23.
[36] Stover, R. C., Sommers, L. E. and Silviera, D. J. (1976), Evaluation of Metals in Wastewater Sludge, *J. Water Pollut. Control Fed.,* **48**, 2165–75.
[37] Stones, T. (1977), Fate of Metals During Sewage Treatment, *Effluent Water Treat.,* 17, 653–55.
[38] Shaheen, D. G. (1975), Contribution of Urban Roadway Usage to Water Pollution, *Environ. Prot. Agency Rep.*, EPA–600/2–75–004.
[39] Pitt, R. (1978), The Potential of Street Cleaning in Reducing Nonpoint Pollution, *Environ. Prot. Agency Rep.*, EPA–600/9–78–017, 91–118.
[40] Pisano, W. C. (1978), Case Study: Best Management Practice (BMP) Solution for a Combined Sewer Problem, *Environ. Prot. Agency Rep.*, EPA–600/9–78–017, 40–52.
[41] Lynard, W. G., Finnemore, P. E. and Finnemore, E. J. (1978), State-of-the-Art of Urban Stormwater Management, *Environ. Prot. Agency Rep.*, EPA-600/9–78–017, 5-21.
[42] Zemansky, G. M. (1974), Removal of Trace Metals During Conventional Water Treatment, *J. Amer. Water Works Assoc.,* **66**, 606–9.
[43] T. Calcutt, (1979), Water Research Centre, private communication.

7

Human exposure to lead and its effects

7.1 Introduction

There is a long history of human exposure to lead, although the relative importance of the different pathways of lead intake may have altered over recent decades. Widespread use was made of lead during the time of the Roman Empire: such uses included the lining of aqueducts and the fabrication of water pipes and cooking utensils, [1]. It is likely that both food and drink will have been substantially contaminated with lead and adverse health effects may have resulted.

Although the toxic effects of metals have merited considerable attention, it should be recognized that many metals also have an essential role to play in biological systems.There is some recent tentative evidence that lead may display this dual role of essentiality and toxicity [2]. Nevertheless, at the concentrations to which the human population is exposed it is reasonable to be concerned principally with the prevention of adverse effects resulting from excessive exposure, rather than the avoidance of possible deficiency syndromes. In this context, it is necessary to have regard to the various sources of human exposure, the intake from these sources, the subsequent uptake and metabolism of the lead in the body and finally, to the adverse effects. These matters will be dealt with separately in the following sections.

7.2 Sources and intake of lead

People are simulaneously exposed in varying degrees to lead in food, drink and air, as well as a number of other sources (Fig. 7.1). There is a continuing debate as to the significance of each of these particular pathways, which is unlikely to be readily resolved, due to the diversity of sources of exposure and their lead concentrations, as well as the varied nature of the individual's metabolic response to the lead. It is, nevertheless, important to attempt to estimate the relative significance of the difference pathways, in order to allow effective implementation of measures designed to limit exposure.

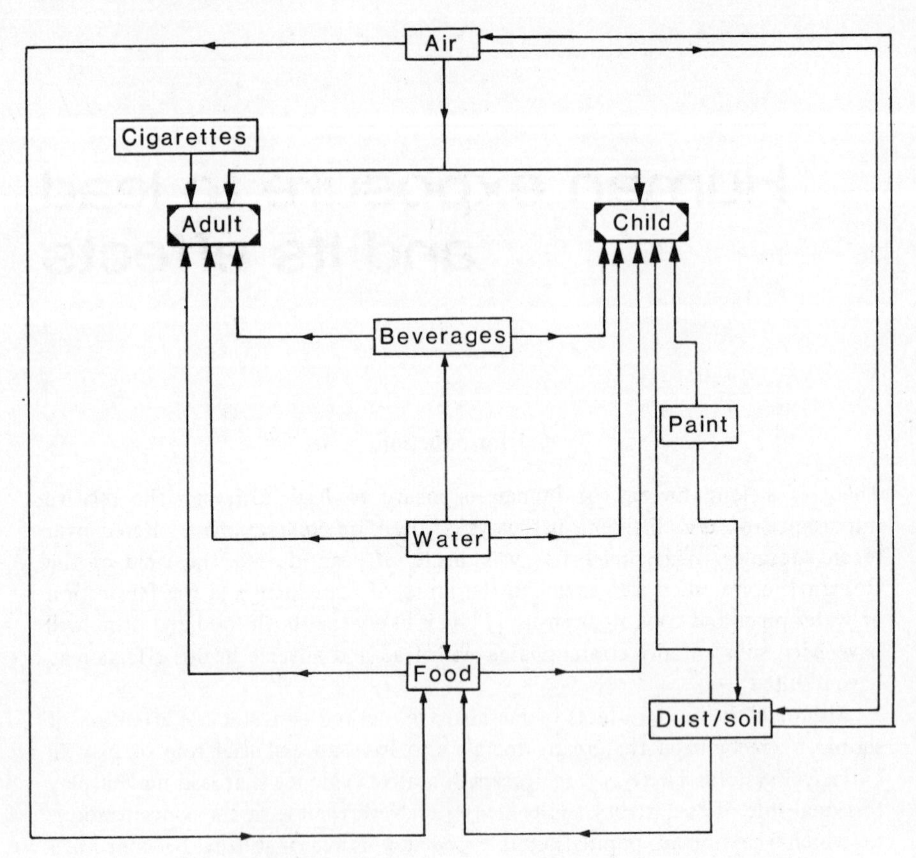

Fig. 7.1 Principal pathways of human exposure to lead.

7.2.1 Food

Food becomes contaminated with lead either at source or during preparation for consumption. Cereals, fruit and vegetables take up lead, albeit inefficiently, from the soil (Section 4.4). They also become contaminated by deposited airborne lead, although a significant part of this may be washed off during preparation for cooking or eating. The extent to which airborne lead contributes to lead in these foodstuffs is unfortunately not established in any fully quantitative manner. Typical concentrations of lead in foods are given in Table 7.1. Lead is also taken up from the water in which vegetables are cooked (Table 7.2). It has been estimated that this additional intake is equivalent to the consumption of an extra 0.18 dm^3 a day of the same tap water [4].

Animals concentrate lead to a certain degree, but most of the lead is deposited in the bone and is hence not available for human consumption. Concentrations

Table 7.1 Lead content of foods in the United Kingdom 1972–74 (based on [3]).

Food type	Mean lead content (mg kg^{-1})*	Range of lead content (mg kg^{-1})
Fresh vegetables		
Brussels Sprouts	0.05	<0.01–0.24
Cabbages	0.09	<0.01–0.54
Carrots	0.04	<0.01–0.12
Potatoes	0.10	<0.01–0.38
Fresh fruit		
Apples	0.26	0.23–0.30
Plums	0.11	0.06–0.16
Meat products		
Cooked	0.20	0.01–0.78
Fresh	0.14	0.05–1.00
Fish (filleted)		
Coastal waters	0.6	<0.2–6.8
Distant waters	0.9	0.4–1.2
Dairy products		
Butter	0.04	0.01–0.09
Eggs	0.09	<0.05–0.35
Cheese	0.12	<0.01–0.18
Milk (fresh)	0.02	<0.02–0.07
Processed foods		
Cereal products	0.21	0.04–0.42
Confectionery	0.14	<0.01–0.45
Canned foods		
Meats	0.75	0.01–2.90
Corned beef	1.20	0.20–6.30
Apples	0.42	0.12–1.14
Blackcurrants	1.20	0.02–7.44
Peas	0.22	0.05–0.72
Tomatoes	0.79	0.16–2.04
Baked Beans	0.30	0.19–0.67
Drinks		
Soft drinks, beer, cider	0.02	<0.01–0.07
Wines	0.12	<0.10–0.72
Canned fruit juices	0.16	0.03–0.59†

*Expressed as mg (Pb) per kg fresh weight of food. In calculating the mean <0.01 was taken equal to 0.01.
†Includes a number of concentrates.

of lead in fresh meat and dairy products are similar to those found in fresh vegetables (Table 7.1). Fish on the other hand accumulate lead to a greater extent and can show much higher concentrations.

The practice of canning foods leads to a considerable enhancement of lead levels, due to the leaching of lead from the soldered seams of cans (Table 7.1).

Table 7.2 Contribution of lead in cooking water to total daily intake of lead from food (based on [4]).

Lead in water (μm dm^{-3})	Daily intake of lead (μg day^{-1})			Lead from cooking water as a percentage of the total
	Vegetables	Other food	Total	
520*	130	64	194	55
265	74		138	37
135	46		110	21
0	23		87	0
250†	67	64	131	40
105	28		92	14
55	23		87	9
0	15		79	0

*Hard water.
†Soft water.

As a result, new regulations were introduced in the UK in 1972 to regulate the concentration of lead in tinned baby foods (Table 7.3). The limit applied to lead in food in Britain since 1979 is 1 mg kg^{-1} with a limit of 0.2 mg kg^{-1} for canned baby foods [4].

With improved food handling practices the lead content of food is generally declining in both Britain [3-5] and the US [6]. The average daily diet includes about 1.5 kg of food. In Britain the Ministry of Agriculture Fisheries and Foods

Table 7.3 Lead content of prepared baby foods in the United Kingdom (based on [3]).

Baby food	Mean lead content (mg kg^{-1})*	Range of lead content (mg kg^{-1})
Canned foods		
(Cans with a lead/tin solder seam, pre-1972)		
Savoury foods	0.12	<0.05–0.53
Sweet foods	0.30	<0.05–0.52
(Cans with a pure tin solder seam, post-1972)		
Savoury foods	0.07	<0.05–0.45
Sweet foods	0.07	<0.05–0.30
Foods in glass jars	0.08	<0.05–0.15
Dehydrated foods		
Milk and cereal products†	0.15	<0.01–0.30

*Expressed as mg (Pb) per kg fresh weight of food.
In calculating the mean <0.05 was taken equal to 0.05.
†In some of these products values are those before reconstitution. The reconstituted values were significantly lower.

Table 7.4 Estimated average daily intake of lead in the United Kingdom, 1972–74, (based on [3]).

Food group	Number of samples	Mean lead content ($mg\ kg^{-1}$)*	Range of lead content ($mg\ kg^{-1}$)	Estimated weight of food eaten (kg)	Estimated lead intake ($\mu g\ day^{-1}$ per person)
Cereals	50	0.12	<0.01–0.70	0.27	32
Meat and fish	50	0.16	<0.01–0.40	0.18	29
Fats	50	0.08	<0.01–0.45	0.08	6
Fruit and preserves	50	0.11	<0.01–0.27	0.25	28
Root vegetables	50	0.09	<0.01–0.70	0.21	19
Green vegetables	50	0.19	<0.01–0.80	0.11	21
Milk	49	0.02	<0.01–0.06	0.40	8
Total	347	0.09†	<0.01–0.80	1.50	140

*In calculating mean <0.01 was taken equal to 0.01

†Weighted according to the proportions of the different food groups consumed.

have simulated a total diet and measured the lead concentration in the various classes of prepared foods. The results, presented in Table 7.4 suggest an estimated average daily intake at that time (1972–4) of 140 μg of lead from foods. Two features of the data should be noted. Firstly, there was an extremely wide range of lead concentrations displayed for each group of foods. Secondly, the mean content of the foods is likely to be biased towards the high side, due to the assumption used in averaging, that concentrations <0.01 mg kg^{-1} were equal to 0.01 mg kg^{-1}.

7.2.2 Drink

Liquids are consumed directly as drinking water, milk and prepared beverages and indirectly in prepared food. The concentrations of lead found in drinking water have been discussed previously (Section 3.3.5). They vary considerably and depend largely on whether the water supply is distributed through lead pipes. Daily consumption of water is typically between 1 and 2 dm^3, which would lead to a daily intake of between 1 and 2000 μg of lead, the values generally being <20 μg day^{-1}.

Lead is present at low concentrations in cows' milk and this source is only likely to contribute 5–20 μg day^{-1} to lead intake. Soft drinks, beer and cider may contribute more significantly to lead intake, (Table 7.1). For instance, a British pint of beer is likely to add about 10 μg to the daily intake of lead, whilst the breakfast drink of fruit juice from a can may add a further 20 μg. More significantly, wine has been found to contain elevated lead concentrations (Table 7.1) and in certain populations it may represent a significant source of lead intake. For instance, the daily consumption of 0.5 dm^3 of wine could lead to a lead intake of *ca.* 60 μg day^{-1}.

7.2.3 Typical daily lead intake from food and drink

The dietary intake of lead has been estimated by a number of methods, the principal ones being firstly the analysis of duplicate portions of meals for total lead and secondly a composite method, whereby typical lead concentrations in the various components of a diet are separately defined and the total lead intake becomes the sum of the lead intakes from the various components, as in Table 7.4. Some of the more recent data, reported since 1970, are summarized in Table 7.5. Clearly there is a high degree of variability in dietary lead intake from individual to individual, as well as from country to country. Typical intakes probably lie in the range 100–200 μg day^{-1} for adults and somewhat lower but not proportionately so, for children at 50–150 μg day^{-1}.

7.2.4 Lead in air

Intake of lead into the body (as distinct from uptake into the bloodstream)

Table 7.5 Dietary intake of lead.

Method	Number of subjects	Age (years)	Sex	Country	Activity	Mean lead intake ($\mu g\ day^{-1}$)	Range of lead intake ($\mu g\ day^{-1}$)	Reference
Duplicate portions	17	Adult	Male	US	Sedentary	113	74–216	[7]
	5	Adult	Male	US	–	308	210–364	[8]
	5	21–30	4 Male 1 Female	UK	–	274	237–306	[7]
	35	Adult	Male	Finland	Medium-heavy	231	119–360	[7]
	36	Adult	Female	Finland	Medium	178	89–305	[7]
	8	0.25–8.5	–	UK	–	–	40–210*	[7]
	195	Adult	–	UK	–	75	21–330	[4]
Composite		18	Male	US	Medium	57–233†	–	[7]
		0.1–1	Male and female	US	–	93±36	–	[9]
		Adult	Male	Canada	Medium	139	–	[7]
		Adult	Male	Italy	Medium	505‡	–	[7]
		All ages	Male and female	UK	–	140	–	[3]
		All ages	Male and female	UK	–	113	–	[4]
		15–17	Male	UK	Medium	146	–	[4]
		2	Male and female	UK	–	70	–	[4]
		3–4	Male and female	UK	–	80	–	[4]

* Lead intake increases with age.
† Varies, depending on the averaging method for lead concentrations reported as zero or trace.
‡ Based on a diet of 3000 calories.

via inhalation will depend upon the daily respired volume and the concentration in the inhaled air. The daily respired volume of air probably lies between 15 and 20 m^3, with the respiration rate being greater during activity than during sleep.

It has already been noted (Section 2.3) that the concentrations of airborne lead will vary considerably, depending amongst other factors on proximity to traffic, whether indoors or outdoors, time of day etc. Chamberlain and co-workers [10] have assessed the concentration data and assume an average daily exposure concentration for urban dwellers in Britain of 1 $\mu g\ m^{-3}$. This concentration would lead to a daily intake of 15-20 $\mu g\ day^{-1}$. Rural concentrations on the other hand are more typically in the range 0.1-0.2 $\mu g\ m^{-3}$, leading to an intake of 1.5-4 $\mu g\ day^{-1}$.

Lead concentrations in excess of 10 $\mu g\ m^{-3}$ may be encountered for short periods of time, for instance in enclosed car parks or inside vehicles on busy urban motorways [10]. These elevated concentrations may lead to a significant enhancement of daily intake.

Not all of the lead in the atmosphere occurs as particulate inorganic lead. Between 1 and 6% of atmospheric lead occurs as tetraalkyllead vapour (Section 2.3). Thus the typical British urban dweller will inhale 0.2-1.2 $\mu g(Pb)\ day^{-1}$ of tetraalkyllead.

7.2.5 Miscellaneous sources

Cigarettes contain from 4 to 12 μg of lead each, a small fraction of which (*ca.* 2%) is transferred to the inhaled smoke. It has been estimated that the intake of lead into the lungs from smoking 20 cigarettes a day is 1-5 μg [7].

Lead in soils, dusts and paints represents a potentially highly important source of lead intake.This is especially so for young children who frequently lick and chew contaminated objects. The source of high concentrations of lead in soils and dusts has been discussed previously (Section 4.3). Paints also exhibit high lead concentrations, for instance primers may contain 30 000-600 000 mg kg^{-1} [5]. Painted surfaces which show high concentrations of lead, in particular those that are flaking, are known to result in enhanced lead intake in certain children. A limit of 2500 mg kg^{-1} is now imposed on lead in paint applied to children's toys, [4] whilst a limit of 5000 mg kg^{-1} is applied in the US on paint used for residential surfaces accessible to children [5].

It is extremely difficult to estimate the amount of lead ingested by children from soils and dusts. Duggan and Williams [11] have produced a crude estimate of a daily consumption by children of 25-100 mg of dirt. For soil and dust lead concentrations ranging from 200 to 2000 mg kg^{-1} this suggests a lead intake of 5-200 $\mu g\ day^{-1}$. Inadvertent ingestion of dirt by a child might therefore double the amount of lead ingested. The above effects are most pronounced in children who exhibit pica, the habit of eating dirt and chewing furniture etc.

7.3 Uptake of lead

The various sources of lead intake, previously discussed, result in uptake via either ingestion of inhalation. However, by no means all the lead taken into the body is absorbed into the bloodstream. Indeed, only a small fraction of ingested lead is absorbed, whilst inhaled lead is absorbed far more effectively.

7.3.1 Ingestion

Gastrointestinal absorption of lead is a relatively inefficient process and most of the lead ingested passes directly to the faeces, to be excreted with no consequence for the human metabolism. Lead isotopes have been used to trace the absorption of ingested lead. They reveal that absorption is highly variable, depending on the chemical form of the lead and whether or not it is ingested with food (Table 7.6). There would also appear to be wide differences in the absorption of lead between individuals exposed under similar conditions. Furthermore, there is some evidence that children absorb lead to a greater extent than adults. The processes regulating absorption have yet to be fully understood.

The general consensus, until recently, has been that the average fractional gastrointestinal absorption of ingested lead is *ca.* 8% [6, 7]. However, Chamberlain and co-workers [10] have related the results of their detailed studies of the metabolism of lead to various data used to derive the above average and conclude that a more realistic average absorption of 13-18% should be applied for adults. These higher values appear necessary to account for the concentrations of lead found in blood and urine.

7.3.2 Inhalation

7.3.2.1 Particulate lead

Lead exists in the atmosphere principally as a fine aerosol of inorganic lead salts (Section 2.5), in a form that is readily inhaled into the alveolar region of the lung. Only a part of the inhaled lead is deposited, the remainder being exhaled. The proportion of lead that is deposited is critically dependent upon the diameter of the lead particles (Fig. 7.2). It also depends on the rate of breathing, the proportion deposited increasing as the breathing becomes slower (Fig. 7.3). For 'Standard Man' the respiratory cycle is assumed to be 15 breaths per minute, resulting in a 24% deposition from an aerosol of 0.09 μm size and 68% for a 0.02 μm aerosol (Fig. 7.3). The average deposition has been found to range from 48 to 64% in a series of inhalation studies using three volunteers in four different environmental settings (Table 7.7).

Virtually all of the deposited lead passes through the alveolar membrane within 2 to 3 days and is redistributed in the circulation (Fig. 7.4). The rate of

Table 7.6 Percentage uptake of lead from the gut.

Lead compound	Percentage uptake from gut*		Reference
	With food	Fasting	
PbS	–	35(3)	[12]
$Pb(NO_3)_2$	9.5(3)	34(3)	
PbS	6±2(6)	12±10(6)	[10]
$PbCl_2$	7±3(6)	45±17(6)	
	Between meals		
$PbCl_2$	21(11)		[13]
$PbCl_2$	20(1)		[10]

*Figures in brackets are the number of subjects.

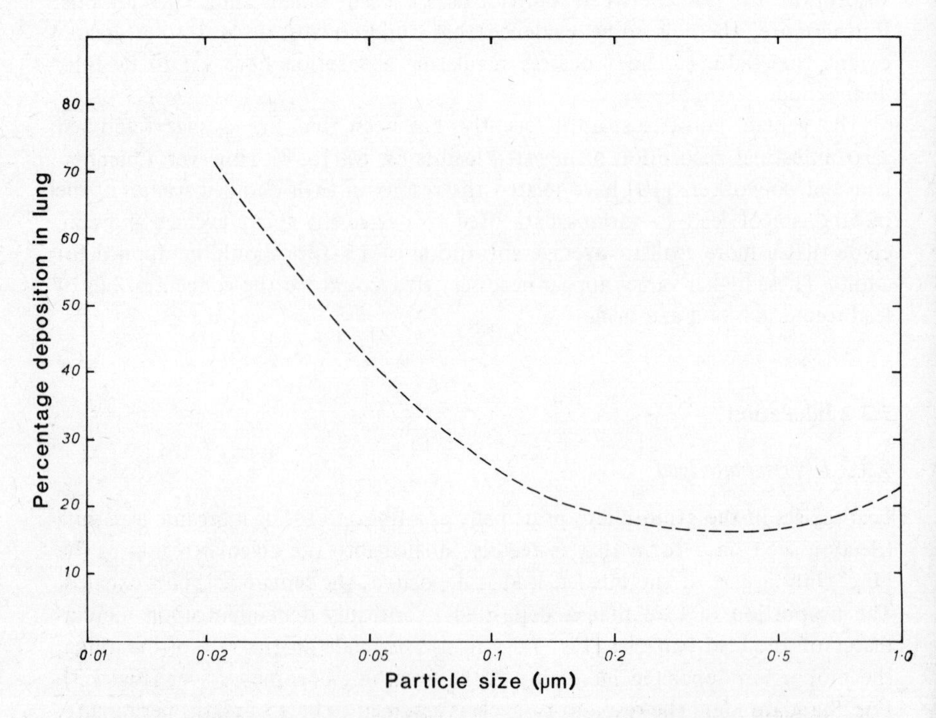

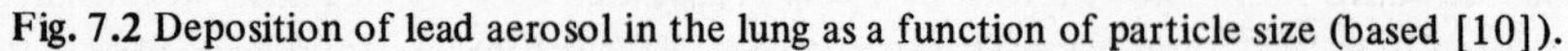
Fig. 7.2 Deposition of lead aerosol in the lung as a function of particle size (based [10]).

clearance is largely unaffected by the chemical form of the lead, except in the case of carbonaceous aerosol which shows a more significant portion with a long half-life. Four exponential terms can be used to describe the clearance. Approximately 22% is cleared with a half-life $T_{1/2}$ of 0.8 h, 34% with a $T_{1/2}$ of 2.5 h, 33% with a $T_{1/2}$ of 9 h and 12% with a $T_{1/2}$ of 44 h.

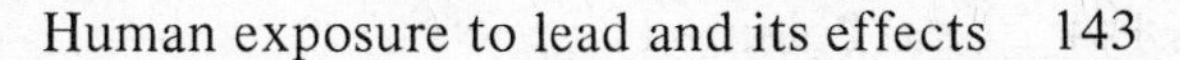

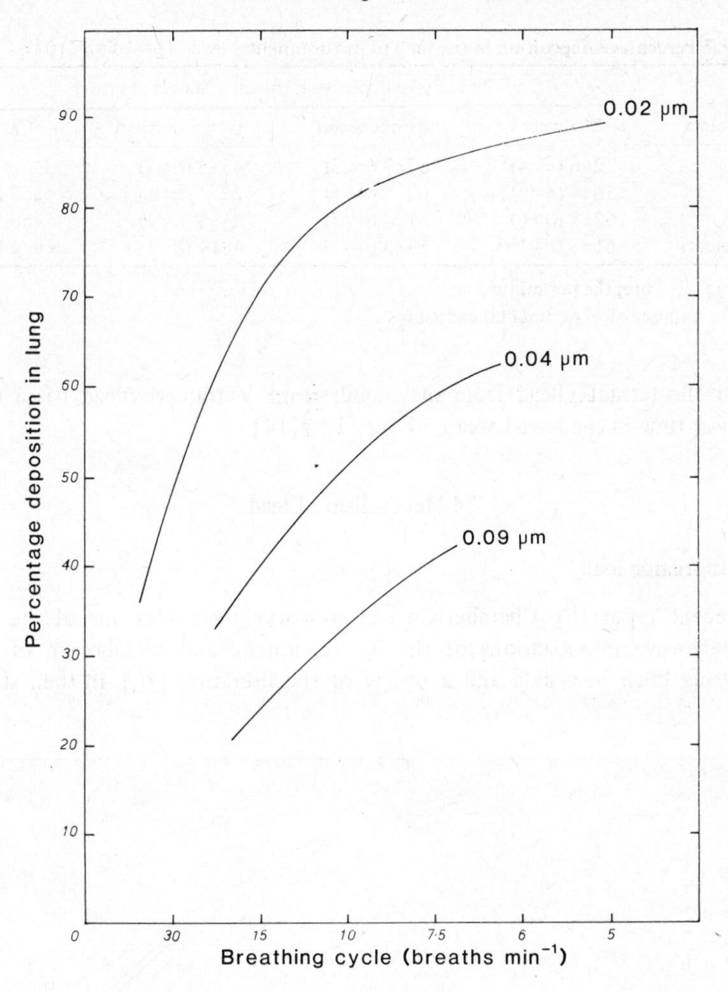

Fig. 7.3 Deposition of lead aerosol in the lung as a function of breathing cycle, at three different aerosol sizes [8].

7.3.2.2 Tetraalkyllead vapour

The uptake of inhaled tetraalkyllead vapour is regulated by a reversible transfer from the air in the lung to the blood as it circulates through the lung.

$$\text{Tetraalkyllead in air} \xrightleftharpoons{\text{lung}} \text{Tetraalkyllead in blood.}$$

The initial uptake of the inhaled vapour has been measured in 3 adult male subjects exposed for 1-4 min to ^{203}Pb labelled tetraalkyllead at 1000 μg m^{-3}. The uptake was *ca.* 50% for tetramethyllead and *ca.* 40% for tetraethyllead [14]. These values probably approximate to the equilibrium uptake, due to the rapid

Table 7.7 Percentage deposition in the lung of environmental lead* (Based on [10]).

Subject	Percentage deposition (mean ± standard error)			
	Motorway	Roundabout	General urban	Car park
A.C.W.	72±6 (n=4)†	67±7 (n=3)	43±5 (n=7)	–
M.J.H.	56±4 (n=9)	62 (n=1)	28 (n=1)	42±3 (n=3)
P.L.	62±5 (n=6)	61±3 (n=3)	54±5 (n=7)	54±6 (n=3)
All Subjects	61±3 (n=19)	64±3 (n=7)	48±4 (n=15)	48±4 (n=6)

*Average 11.3 breaths per minute.
†n is the number of experimental exposures.

loss of the tetraalkyllead from the bloodstream. Tetramethyllead has a mean residence time in the bloodstream of only 13 s [14].

7.4 Metabolism of lead

7.4.1 Inorganic lead

The recent report by Chamberlain and co-workers provides one of the most comprehensive investigations of the intake, uptake and metabolism of lead, presenting both new data and a review of the literature [10]. In their studies

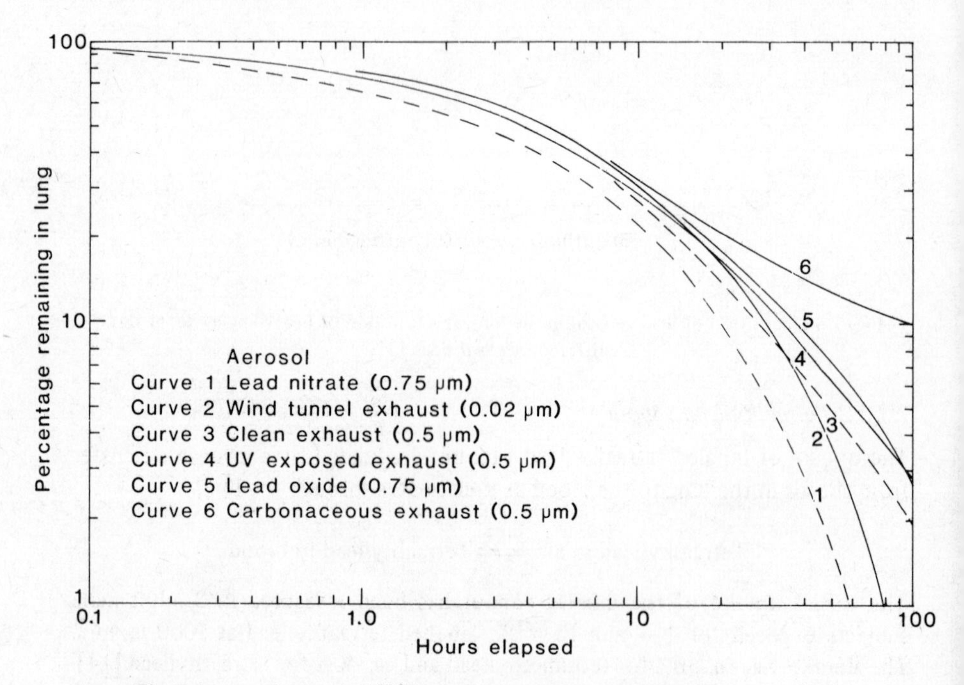

Fig. 7.4 Lung clearance of ^{203}Pb labelled aerosols deposited in the lung [10].

on the metabolism of lead they used ^{203}Pb to trace the absorbed lead, for a period of up to ten days. The behaviour of the lead, once absorbed, was found to be independent of whether uptake was by inhalation or ingestion.

Lead absorbed into the body enters the bloodstream initially, and rapidly attaches itself to the red blood cells. There is a further rapid redistribution of the lead between blood, extracellular fluid and other storage sites, such that only about half of the freshly absorbed lead is found in the blood after only a few minutes. The storage site for the lead is uncertain, although it is probably in soft tissue, as well as bone. Indeed, the lead stored in bones represents the major body burden of lead in humans. There appears to be a steady accumulation of lead in bones throughout life (Fig. 7.5), the amount entering long-term storage increasing with the uptake of lead (Fig. 7.6). There is, however, evidence that not all of this lead is immobile. Some will be continuously re-entering the bloodstream, resulting in a half-life for this lead in bone of between 600 and 3000 days. Furthermore, there is some ill-defined evidence that bone lead may be remobilized more rapidly when the body is suffering certain conditions of stress [5], although objective evidence to confirm this possibility is lacking.

The lead that remains in the bloodstream is lost steadily with a short-term half-life of *ca.* 20 days (adult males, normal blood lead concentration). The half-life for excretion of the lead from the body is, however, somewhat longer at *ca.* 28 days [10]. This latter figure is consistent with the half-life of lead in the bloodstream found in the long-term studies of Rabinowitz [8], which suggests that some of the lead initially lost to storage re-enters the bloodstream before being excreted.

There is now considerable evidence to show that the half-life for excretion of lead from the body is not constant, but is a function of the blood lead concentration (henceforth, blood lead concentration will be called simply blood lead or PbB). As the blood lead increases, so the half-life of lead in the body decreases. There is thus a certain natural protective mechanism against exposure to increased amounts of lead. The evidence for this takes the form of an increased rate of urinary excretion of lead, the major pathway for removal of absorbed lead, per unit of blood lead, as PbB increases. This observation is also consistent with the numerous epidemiological studies which approximate the relationship of blood lead levels to exposure by a log–linear or log–log plot of a slope <1, i.e. one in which the incremental increase in blood lead per unit of intake decreases as the intake increases. It may also be that the fractional uptake of lead into the bloodstream decreases as the body is exposed to higher concentrations.

The uptake and metabolism of lead is summarized in Fig. 7.7. This model represents a synthesis of the information presented by Chamberlain *et al.* [10] and Rabinowitz *et al.* [8, 15]. It must, nevertheless, remain somewhat speculative at the present time, although it is numerically self-consistent. Furthermore, it represents 'typical' conditions, at least in so far as one can typify

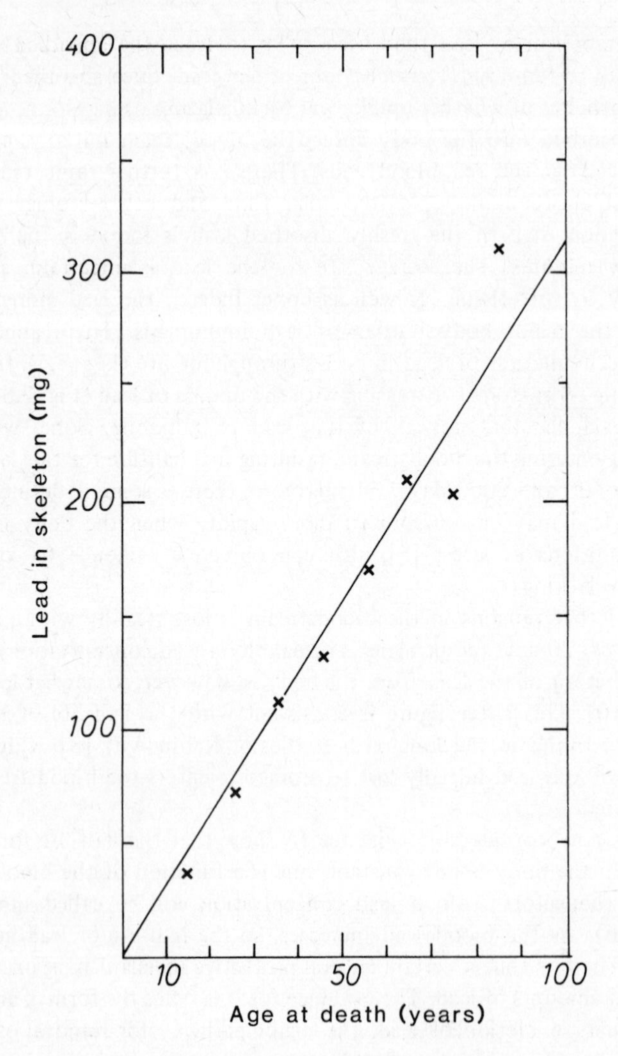

Fig. 7.5 Lead in the skeleton at death, normalized to 70 kg body weight [10].

the highly variable nature of each of the stages of intake, uptake and metabolism (Sections 7.2, 7.3 and 7.4). The assumptions involved in deriving this model are that it represents uptake by a male, *ca.* 30 years of age, with a blood volume of 5 dm^3, eating a representative diet (Section 7.2.3) and breathing air in an urban setting (Section 7.2.4). Under these conditions the uptake of lead from the atmosphere (1 μg (Pb) m^{-3}) represents 29% of the total uptake. The predicted blood lead is within the range of averages recently measured (1979) for adults in British cities [4].

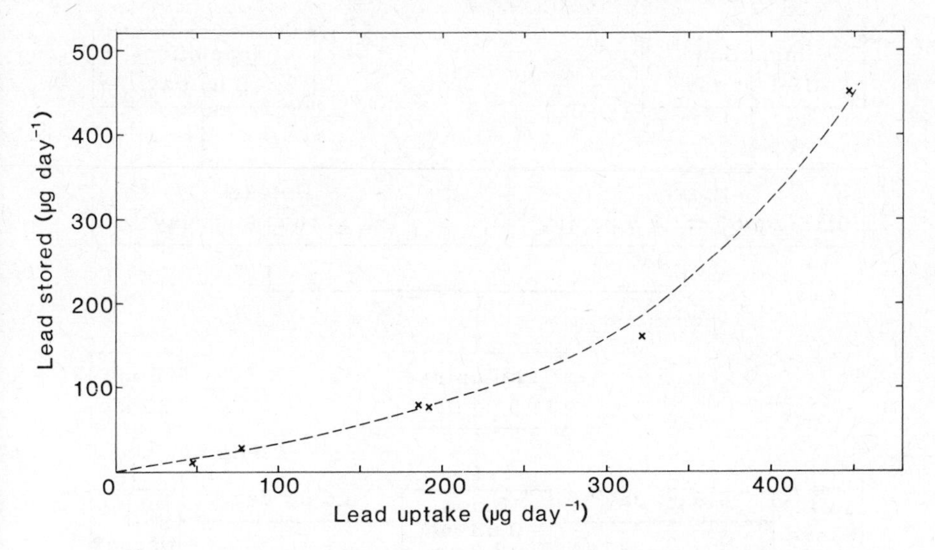

Fig. 7.6 Amount of lead in long-term storage as a function of lead uptake. Based on data of Kehoe, cited in [10], with uptake calculated from gut absorption = 0.15 × intake.

7.4.2 Tetraalkyllead

The metabolism of tetraalkyllead is different from that of inorganic lead. Initial entry is still into the bloodstream but the lead is more evenly distributed between blood plasma and red blood cells. Tetraalkyllead, in contrast to inorganic lead, is lost rapidly from the bloodstream, although after 5 to 10 h some of the lead re-appears, associated this time almost exclusively with the red blood cells [14]. The lead is now probably present as trialkyllead, with some possibly having been converted to inorganic lead.

Organic lead is found to concentrate in the liver and it is probably here that the tetraalkyllead is converted to the trialkyllead form [16]. Otherwise the lead is widely dispersed about the body. It is excreted only slowly, with a half-life of the order of 200 to 350 days [14].

7.5 Blood leads

The lead circulating in the bloodstream is mobile, in contrast to that stored in bones, and it is this lead that is able to exert adverse effects on the body. Hence the concentration of lead in the bloodstream (PbB) is an important parameter in the characterization of an individual's exposure to lead and, more importantly, in the relationship between exposure and the occurrence of adverse consequences. A major proviso, however, is that blood leads represent only recent exposure to lead. They do not reflect previous exposures which may have been considerably higher. This proviso must detract from some of the correlations

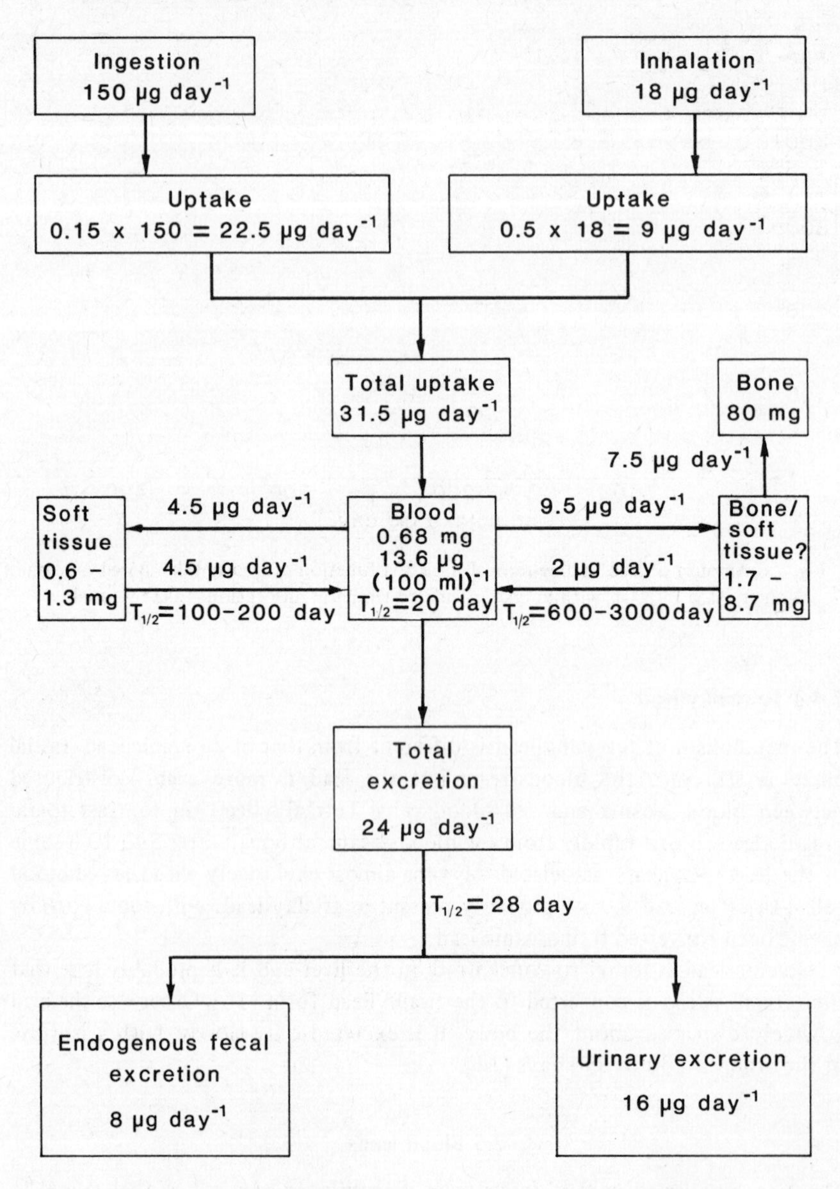

Fig. 7.7 Tentative model of metabolism of lead in healthy adult males. (See text for further details, Section 7.4.1).

found between existing blood leads and adverse effects, which may in fact relate to previous much higher blood lead conditions. Naturally, this should lead to caution in defining and interpreting definite dose–response relationships. One way around this problem, adopted by some workers, has been to relate effects to the lead concentration in teeth (i.e. bone lead) which will better reflect the integrated history of exposure to lead.

A further problem relating to the use of blood leads is that of the accuracy of the reported values. Extreme care is required in the sampling and analysis in order to obtain reliable results. Studies have revealed wide discrepancies between results obtained in different laboratories, which must caution against the uncritical acceptance of reported blood leads, especially at the lower blood lead levels [17]. This would also seem to apply to measurements of tooth lead concentrations. Greater care is now exercised and workers often report their efforts to establish accuracy through participation in interlaboratory schemes.

A range of indices, in addition to blood leads, have become available to measure exposure to lead. These include blood measurements of δ-aminolaevulinic acid dehydratase (δ-ALAD) or free erythrocyte protoporphyrin (FEP) [18]. The wealth of data currently available refer, nevertheless, to blood lead values.

7.5.1 Population values

Blood leads in the general population are highly variable. This is hardly surprising given the high degree of variability in the intake, uptake and metabolism of lead (see previous sections). The mean blood lead values for different population groups generally lie in the range 10–30 $\mu g\ (100\ ml)^{-1}$ for adults and are usually marginally higher for children [6]. Perhaps somewhat surprisingly this range of means is applicable to most countries of the world, as well as to a variety of environmental settings [6].

The population mean blood lead disguises a wide distribution of individual blood leads. These values are generally found to be log-normally distributed and most data point to a geometric standard deviation of 1.3 about the geometric mean [6]. The effect of this distribution centred on a population geometric mean PbB of 15 $\mu g\ (100\ ml)^{-1}$ is illustrated in Fig. 7.8. It is evident that 5% of this population would have a PbB $> 23\ \mu g\ (100\ ml)^{-1}$, whilst 1% would be expected to have a PbB $> 28\ \mu g\ (100\ ml)^{-1}$ and 0.01% $> 40\ \mu g\ (100\ ml)^{-1}$. For instance, in a population of 1 million, 100 people would be expected to have blood leads in excess of 40 $\mu g\ (100\ ml)^{-1}$ if the population mean were 15 $\mu g\ (100\ ml)^{-1}$.

The implication of this wide range of blood leads within a population is important for standard setting based on PbB. For instance, the US Environmental Protection Agency (US EPA) has promulgated a standard for lead in air based on achieving a target of a population median PbB for children of $<15\ \mu g\ (100\ ml)^{-1}$ [19]. They are thus implicitly accepting that 1 in every 10 000 children may have a blood lead $>40\ \mu g\ (100\ ml)^{-1}$. This standard is, nevertheless, effectively the same as one based on an individual PbB limit of 35–50 $\mu g\ (100\ ml)^{-1}$ as applied in some countries.

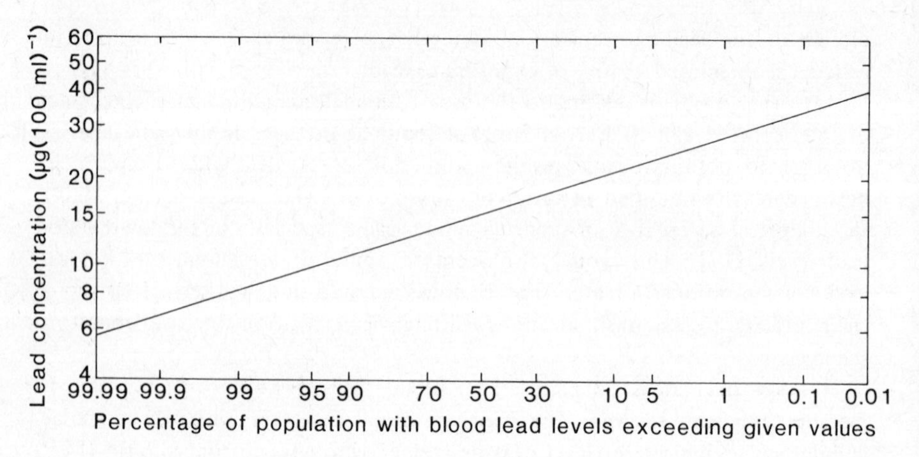

Fig. 7.8 Distribution of blood lead values with geometric mean of 15 μg (100 ml)$^{-1}$ and geometric standard deviation of 1.3.

7.5.2 Relationship to source of intake

One issue that has demanded considerable attention over recent years has been the contribution of vehicle-generated lead aerosol to human uptake of lead and hence blood lead levels. There are three principal pathways of human exposure to this source of lead; direct inhalation, ingestion of deposited dust (street dust, household dust, soil) and ingestion of foods contaminated directly or indirectly by deposited lead.

Of these three pathways, there is at present no available information to assess the magnitude of the contribution of deposited atmospheric lead to the third pathway. There is on the other hand considerable information upon the incremental change in blood lead (dPbB) caused by a small increase in air lead concentration (dPbA) i.e. the ratio dPbB/dPbA (Table 7.8). It is important to remember when examining these data that the ratio dPbB/dPbA is likely to decrease as the exposure to lead increases. There is, furthermore, some evidence that the ratio dPbB/dPbA is higher in children than adults [6].

There is less certainty about the magnitude of the effect of the direct ingestion of lead in deposited dust (street dust or dust incorporated in soil). The possible magnitude of this intake by young children, who are particularly susceptible, has already been discussed (Section 7.2.5). There are in addition some epidemiological data which are summarized in Table 7.9. These data show a remarkable consistency when analysed as a log-log relationship and when applied at comparable exposure levels. They would suggest an increase in PbB for children of *ca.* 3 μg (100 ml)$^{-1}$ when the soil lead (PbS) increases from 200 to 1200 mg kg^{-1}. It must be noted however that these studies do not prove that the soil lead was the sole cause of the increase in blood lead levels.

Table 7.8 Summary of incremental blood lead/air lead ratios, dPbB/dPbA.

Study		Ratio dPbB/dPbA
Epidemiological studies		range 0.11–3.84, mean 1.5 (m=13)*
Clinical studies		
Kehoe [20, 21]	– adult male (n=3)* at 10–40 $\mu g\ m^{-3}$	0.80
Griffin [22]	– adult male (n=8) at 10.9 $\mu g\ m^{-3}$	1.45
Griffin [22]	– adult male (n=13) at 3.2 $\mu g\ m^{-3}$	1.6
Kehoe [20, 21]	– adult male (n=2) at 0.6–4.4 $\mu g\ m^{-3}$	1.4
Rabinowitz [15]	– adult male (n=4) at 2 $\mu g\ m^{-3}$	3.5
Chamberlain [10]†	– adult male at 1 $\mu g\ m^{-3}$	2.6

* n is the number of experimental exposures and m is the number of studies.

† Based on model calculations, allowing for long-term resorption from storage sites.

Table 7.9 Blood leads of young children as a function of soil lead (based on [6]).*

Study location	Children's age (years)	Geometric mean blood lead (μg (100 ml)$^{-1}$)	Geometric mean soil lead (mg kg^{-1})	dPbB/dPbS at geometric mean soil Pb (μg (100 ml)$^{-1}$ / 1000 mg kg^{-1})	dPbB/dPbS at 1000 mg kg^{-1} soil Pb (μg (100 ml)$^{-1}$/ 1000 mg kg^{-1})	Increase in PbB for a soil Pb. Increase from 200 to 1200 mg kg^{-1} (μg (100 ml)$^{-1}$)
Charleston, South Carolina	1–5	36.4	452	3.5	1.6	2.8
Kellogg, Idaho	1–9	37.5	1518	1.3 (0.6)†	1.9 (0.6)†	3.3 (0.6)†
Dallas, Texas	1–5	11.4	92	8.2	0.9	1.5
Derybshire, England	2–3	23.2	1849	1.0	1.9	3.1

*The US EPA re-analysed the original data using a log–log relationship.

†The value cited in the original study used a log (blood lead) – linear (soil lead) relationship. This is an inappropriate relationship for blood lead-exposure studies (Section 7.5.2).

7.6 Biological and adverse health effects of exposure to lead

In the previous sections an attempt has been made to define the major pathways of human exposure to lead and then to examine the resultant blood lead concentration as a biological indicator of this exposure. It thus follows that the biological effects should, in so far as possible, be likewise related to this same index of exposure, the blood lead concentration. Furthermore, it is this mobile lead in the bloodstream that is the cause of the adverse effects of lead.

A problem inherent in the study of lead poisoning is that the symptoms are largely non-specific (Table 7.10). This can lead to difficulties in the diagnosis of cases of over-exposure to lead. The first consequence of exposure to lead to be considered is its effect on the blood itself.

Table 7.10 Symptoms of inorganic and organic lead poisoning [5].

Inorganic		Organic
Adults	Children	
Abdominal pain	Drowsiness	Disturbance in sleep pattern
Constipation	Irritability	Nausea
Vomiting	Vomiting	Anorexia
Non-abdominal pain	Gastrointestinal symptoms	Vomiting
		Vertigo and headache
Asthenia	Ataxia	Muscular weakness
Paraesthesiae	Stupor	Weight loss
Psychological symptoms	Fatigue	Tremor
		Diarrhoea
Diarrhoea		Abdominal pain
		Hyperexcitability
		Mania

Symptoms are listed in their order of frequency as presenting symptoms.

7.6.1 Bloodstream

The best documented effects of lead on blood are its interference with the biosynthesis of haem, which is essential for the production of haemoglobin (the red oxygen-carrying pigment in red blood cells). The two most important points of interference are the formation of the enzyme δ-aminolaevulinic acid dehydratase (δ-ALAD) and the insertion of iron into protoporphyrin [6].

Inhibition of δ-ALAD is first noted at blood lead levels of 10–20 $\mu g\ (100\ ml)^{-1}$, with children more susceptible than adults. It seems unlikely however, that haem synthesis is inhibited until a 20–30% decrease in δ-ALAD has occurred [6].

Lead interference with the insertion of iron into protoporphyrin results in an accumulation of protoporphyrin in erythrocytes (red blood cells). It appears that free erythrocyte protoporphyrin (FEP) begins to increase at a PbB of 17-20 $\mu g\ (100\ ml)^{-1}$ in children and women, and at slightly higher levels of 20–25 $\mu g\ (100\ ml)^{-1}$ in adult males [6], although these thresholds are ill-defined and are still open to question. This accumulation of FEP has been taken by the

US EPA to be the first sign of physiological impairment due to lead in humans [19].

A summary of threshold levels for the various effects of lead in humans, based on a thorough survey of the literature by the US EPA, prior to their setting of a standard for airborne lead, is given in Table 7.11.

7.6.2 Neurobehavioural effects

The most profound effects of lead poisoning are undoubtedly those associated with severe damage to the central nervous system. At high levels of exposure to lead, neural (brain) damage may result in stupor, convulsions and/or coma and may progress to death. It is also generally accepted that severe neural damage due to lead poisoning may be irreversible. It is precisely this irreversibility, which, if it also applies to the more subtle neural effects of lead, has given rise to the concern about the exposure of young children, who are especially susceptible to lead in the environment.

Unfortunately, there is no general agreement as to the threshold lead levels in blood that cause obvious central nervous system symptoms. Lead encephalopathy, a degenerative disease of the brain, is, however, unlikely to occur in adults with PbB $<120\ \mu g\ (100\ ml)^{-1}$. The evidence points to a lower threshold of $100\ \mu g\ (100\ ml)^{-1}$ for children [6].

Children are acknowledged to be more susceptible to lead poisoning for a wide variety of reasons, which include a greater intake and hence uptake of lead from the diet and the atmosphere per unit body weight, as well as the incomplete development of metabolic pathways and the blood-brain barrier [6]. Concern is therefore, now focused on the neurobehavioural effects of lead in children, in particular on the more subtle effects, which have become widely known as 'sub-clinical', although this term is strictly self-contradictory. These effects consist mainly of cognitive (mental) or sensory-motor integration (co-ordination) deficiencies.

Subtle lead poisoning represents a particularly controversial area of research at present. Studies with similarly exposed groups have shown both positive and negative relationships between lead exposure and neurobehavioural effects. Many studies, however, reveal methodological faults and weaknesses, which limit the validity of some of the conclusions. Nevertheless, the study, by Needleman and co-workers [23], is generally accepted as providing the most reliable information currently available [24]. As this is such an important study, it will be considered in some detail.

7.6.2.1 The Needleman study

This study took as its sample over 3000 children, aged 7-8 years, at school in Chelsea and Somerville, Massachusetts. The index of lead exposure was the tooth dentine lead concentration. Teacher's ratings on 11 aspects of classroom behaviour

Table 7.11 Lowest observed effect levels (based on [6]).

Lowest observed effect level ($\mu g\ (100\ ml)^{-1}$)	Effect	Population group
10	δ-ALAD enzyme inhibition	Children and adults
17–20	Erythrocyte protoporphyrin elevation	Women and children
25–30	Erythrocyte protoporphyrin elevation	Adult males
40	Increased urinary ALA excretion	Children and adults
40	Anaemia	Children
40	Co-proporphyrin elevation	Children and adults
50	Anaemia	Adults
50–60	Cognitive (central nervous system) deficits	Children
50–60	Peripheral neuropathies	Children and adults
80–100	Encephalopathic symptoms	Children
100–120	Encephalopathic symptoms	Adults

of 2146 individuals who submitted teeth for analysis were subsequently related to the tooth dentine lead concentration (the teacher was blind to the lead level). Some of the results are shown in Fig. 7.9. One result of interest is that hyper-

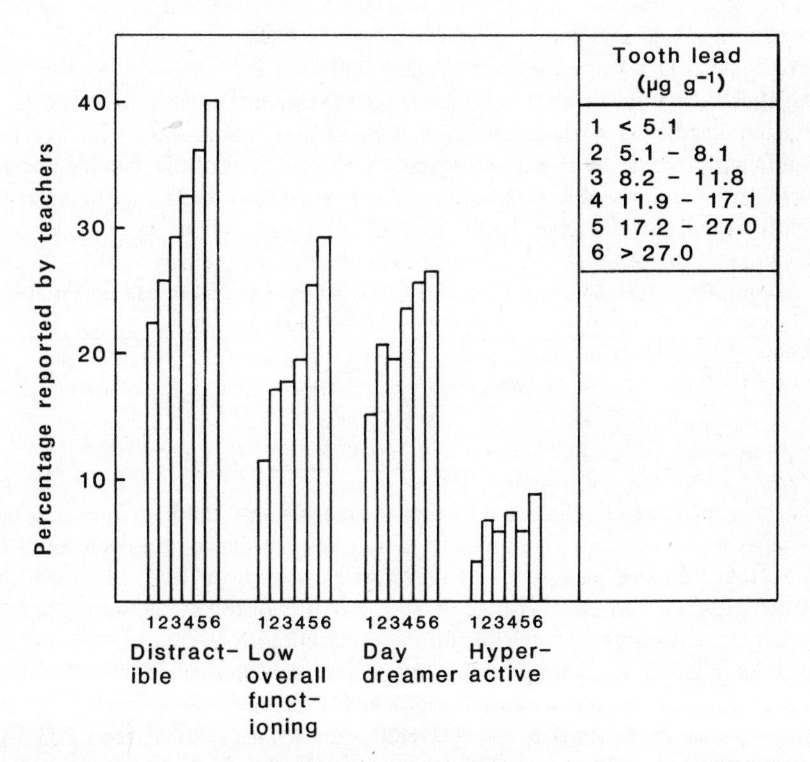

Fig. 7.9 Teachers negative ratings on classroom behaviour in relation to tooth lead concentration (based on [23]).

activity did not appear to relate to lead level. This is contrary to the results of others who have suggested this to be a symptom of subtle lead poisoning [24]. The various classifications of classroom behaviour generally show a consistent dose-response relationship, which strengthens the argument for a causal relationship.

This rather crude analysis does not however take into account other factors which may cause an apparent relationship. For instance classroom performance can be related to parental background, which in turn might result in different exposures to lead. To allow for possible confounding variables, the study examined a smaller number of children in greater detail.

For this exercise, children with high tooth leads, >20 μg g^{-1}, were compared with those with low tooth leads, <10 μg g^{-1} (agreement between replicate tooth samples was required for inclusion in one or other of the two groups). The children, 100 in the low lead group and 58 in the high, were then subjected to a battery of neuropsychological tests. In addition, parental factors were also assessed. There was a significant relationship between the child's tooth lead and several parental factors, including parental education and father's socio-economic status. These factors were subsequently allowed for in an analysis of co-variance.

The test scores, after allowing for parental factors showed a general tendency for the high-lead children to perform less well. The mean full scale IQs were 106.6 and 102.1 respectively in the low and high lead children, a difference statistically significant at the 3% level. One factor not used as a co-variate in this initial analysis was a childhood history of pica, which was found to relate significantly to the tooth lead grouping. Further analysis of the data [24] would suggest that it is possible that both pica and raised lead levels may be independently associated with lower IQ (Table 7.12).

Table 7.12 Mean values of performance IQ as a function of tooth lead group and history of pica (based on [24]).

	IQ with no history of pica	IQ with a positive history of pica	IQ for all children
High lead	106.0	103.4	104.9
Low lead	110.7	106.9	108.7

The significance of this study is that it takes normal children from an urban setting and finds what would appear to be strong evidence of a decrease in IQ and other functioning due to the integrated exposure to lead, as defined by tooth leads. The problem with applying the results of this study more generally is that there is insufficient reliable information on tooth lead concentrations and on their relation to exposure levels. Blood lead determinations were, however, available for about half of the subjects, taken four to five years previous to the study. Those in the high tooth lead group had a mean blood lead of 35.5 μg (100 ml)$^{-1}$ (range 18–54 μg (100 ml)$^{-1}$), whilst those in the low lead group had a mean of 23.8 μg (100 ml)$^{-1}$ (range 12–37 μg (100 ml)$^{-1}$). These are both high mean values by normal standards and it would seem that even those in the low lead group had experienced greater than average exposure to lead at some time in the past.

The Needleman study is not without its limitations and it is essential that further careful studies be carried out to confirm, modify or reject the findings [24]. Furthermore a number of questions require answering. For instance, if the damage is irreversible, is there a threshold blood lead for this effect? If so, how does this affect the use of tooth leads as an index of exposure to previous harmful blood leads? Two patterns of blood leads producing identical tooth lead concentrations might be quite different.

7.6.2.2 Conclusions

The literature relating to subtle neurobehavioural impairment due to lead has been critically reviewed up to the year 1977 by the US EPA [6]. They concluded that adverse effects will be seen in some children at blood lead levels of 50 to 60 $\mu g\ (100\ ml)^{-1}$. A more recent critical review by Rutter [24], for the British working party on Lead and Health [4], concluded that it seems very likely that intellectual deficits may occur in individuals with blood leads in the range 40 to 80 $\mu g\ (100\ ml)^{-1}$ during exposure. A more important question, however, as addressed by the Needleman study, is whether there can be cognitive impairment in individuals who have never had severe lead exposure but whose body burden of lead is somewhat above the average for the general population.

Amidst all this uncertainty the US EPA has set a standard for lead in air of 1.5 $\mu g\ m^{-3}$ based on achieving a target of a population median blood lead for young children (1-5 years) of $<15\ \mu g\ (100\ ml)^{-1}$. This median has been chosen in order to prevent most children in the US from exceeding a blood lead of 30 μg $(100\ ml)^{-1}$ (see Fig. 7.8). This value has been adopted as a maximum safe blood lead for individual children and is based principally upon the elevation of FEP in the blood as an index of adverse biochemical effect [19]. By adopting this approach they have avoided basing a standard on the uncertain neurobehavioural blood lead relationships.

References

[1] Grandjean, P. (1975), Lead in Danes: Historical and Toxicological Studies, in *Environmental Quality and Safety, Supplement Vol. II*, (ed. F. Coulston and F. Corte), Academic Press, New York, San Francisco, London, pp. 6–75.

[2] Schwarz, K. (1977), Essentiality versus Toxicity of Metals, in *Clinical Chemistry and Chemical Toxicilogy of Metals*, (ed. S. S. Brown), Elsevier/North-Holland Biomedical Press, Amsterdam, pp. 3–22.

[3] Ministry of Agriculture, Fisheries and Food (1975), Working Party on the Monitoring of Foodstuffs for Heavy Metals, 5th Report, Survey of Lead in Food: 1st Supplement Report, HMSO, London, pp. 34.

[4] Department of Health and Social Security (1980), *Lead and Health*, Dept. Health and Social Security, HMSO, London, p. 129.

[5] Waldron, H. A., Stöfen, D. (1974), *Sub-Clinical Lead Poisoning*, Academic Press, London, New York, p. 224.

[6] US Environmental Protection Agency (1977), Air Quality Criteria for Lead, *EPA Rep. EPA-600/8-77-017.*

[7] World Health Organisation (1977), *Environmental Health Criteria, 3, Lead*, WHO, Geneva, p. 160.

[8] Rabinowitz, M. B., Wetherill, G. W. and Kopple, J. D. (1976), Kinetic Analysis of Lead Metabolism in Healthy Humans, *J. Clin. Invest.*, 58, 260–70.

[9] Kolbye, A. C., Mahaffey, R., Fiorino, A., Corneliussen, P. C. and Jelinek, C. F. (1974), Food Exposure to Lead, *Environ. Health, Perspect., Exp. Issue* 7, 65–75.

[10] Chamberlain, A. C., Heard, M. J., Little, P., Newton, D., Wells, A. C. and Wiffen, R. D. (1978), *Investigations into Lead from Motor Vehicles*, HMSO, London.

[11] Duggan, M. J. and Williams, S., (1977), Lead-in-Dust in City Streets, *Sci. Tot. Environ.*, 7, 91–97.

[12] Wetherill, G. W., Rabinowitz, M. and Kopple, J. D. (1975), Sources and Metabolic Pathways of Lead in Normal Humans, *Proceedings of the Conference on Recent Advances in the Assessment of Health Effects of Environmental Pollution*, Paris, 1974, Commission of European Communities, Luxembourg, pp. 847–60.

[13] Blake, K. C. H. (1976), Absorption of ^{203}Pb from the Gastrointestinal Tract of Man, *Environ. Res.*, **11**, 1–4.

[14] Heard, M. J., Wells, A. C., Newton, D. and Chamberlain, A. C. (1979), Human Uptake and Metabolism of Tetra Ethyl and Tetra Methyl Lead Vapour Labelled with ^{203}Pb, *Proceedings of the International Conference of Heavy Metals in the Environment*, London, Sept. 1979, CEP Consultants Ltd., Edinburgh, pp. 103–108.

[15] Rabinowitz, M. B., Wetherill, G. W. and Kopple, J. D. (1977), Magnitude of Lead Intake from Respiration by Normal Man, *J. Lab. Clin. Met.*, 90, 238–48.

[16] Grandjean, P. and Nielsen, T. (1979), Organolead Compounds: Environmental Health Aspects, *Residue Rev.*, 72, 97–148.

[17] Pierce, J. O., Koirtyohann, S. R., Clevenger, T. E. and Lichte, F. E. (1976), *The Determination of Lead in Blood. A Review and Critique of the State of the Art 1975*, International Lead and Zinc Research Organization, New York.

[18] Roots, L. M. (1979), Tests Available for Assessing Recent Exposure to Inorganic Lead Compounds and Their Use for Screening Purposes. *Sci. Tot. Environ.*, **11**, 59–68.

[19] US Environmental Protection Agency (1978), National Ambient Air Quality Standard for Lead, *Federal Register*, **43**, 46246–77.

[20] Kehoe, R. A. (1961), The Metabolism of Lead under Abnormal Conditions, *J. Roy. Inst. Public Health*, **24**, 101–43.

[21] Kehoe, R. A. (1966), Criteria for Human Safety from the Contamination of the Ambient Atmosphere with Lead, *Proceedings of the International Congress on Occupational Health*, Vienna, pp. 83–98.

[22] Griffin, T. B., Coulston, F., Wells, H., Russell, J. C. and Knelson, J. H. (1975), Clinical Studies on Man Continuously Exposed to Airborne Particulate Lead, in *Lead*, (ed. T. B. Griffin and J. H. Knelson), Academic Press, New York, pp. 221–40.

[23] Needleman, H. L., Gunnoe, C., Leviton, A., Reed, R., Peresie, H., Maher, C. and Barrett, P. (1979), Deficits in Psychologic and Classroom Performance of Children with Elevated Dentine Lead Levels, *New England J. Med.*, **300**, 689–695.

[24] Rutter, M., (1980), Raised Lead Levels and Impaired Cognitive/Behavioural Functioning: A Review of the Evidence, *Suppl. Develop. Med. Child Neurology*, **22**, 1–26.

8

Chemical analysis of lead in the environment

8.1 Introduction

It will be clear from the preceding chapters that reliable analysis of lead in industrial and environmental samples is an essential prerequisite to the assessment of human exposure, the understanding of environmental pathways of the metal and the evaluation of the effectiveness of pollution control technologies. Consequently a brief summary of the more important considerations in the chemical analysis of lead is pertinent.

8.2 Chemical analysis of lead: available techniques

There are many analytical techniques which may be used for lead analysis. These include: X-ray fluorescence (XRF) spectroscopy, radioactivation methods, emission spectrography, ring oven methods, polarographic techniques [including anodic stripping voltammetry (ASV)], spark source mass spectrometry, colorimetry and atomic absorption spectrometry (AAS). Background information upon all of these methods may be found in the 'Handbook of Air Pollution Analysis' [1].

The choice of analytical method is obviously largely dependent upon the availability of instrumentation. There are four techniques, however, which are used far more widely in the analysis of lead in environmental samples than other methods: they are XRF, ASV, colorimetry with dithizone and AAS. Because of its versatility, ease of use and the low capital cost of equipment, atomic absorption is by far the most commonly used technique. Comments will be restricted to these four more important methods. Detection limits based upon experience, rather than manufacturer's literature are cited in Table 8.1. Obviously these may be an important determinant of the techniques selected for low-level work.

8.2.1 X-Ray fluorescence (XRF)

The main disadvantage of XRF is the high initial capital cost of instrumentation. The method is used, however, particularly in the analysis of particulate lead in

Table 8.1 Typical detection limits in lead analysis.

Technique	Detection limit
X-ray fluorescence	0.2 $\mu g\ cm^{-2}$ *
Dithizone colorimetry	20 $\mu g\ dm^{-3}$ †
Anodic stripping voltammetry	0.01 $\mu g\ dm^{-3}$ ‡
Atomic absorption – flame	10 $\mu g\ dm^{-3}$ §
Atomic absorption – flameless	0.02 $\mu g\ dm^{-3}$ ¶

* For analysis of the surface of an air filter by the thin film technique.
† Based upon extraction of a 10 ml aliquot of sample solution.
‡ With a hanging mercury drop electrode. Can be improved with alternative electrodes.
§ With a double beam instrument.
¶ Based upon injection of a 100 μl aliquot of solution.

air. The advantage of XRF in this application is that by using the thin film technique, atmospheric particles collected on a filter may be examined by direct insertion of the air filter into the instrument without pretreatment. Sensitivity is not especially high, and rather lengthy sampling times may be required to collect sufficient lead for analysis. For routine analysis of large numbers of 24 h air samples XRF may be the preferred technique.

8.2.2 Anodic stripping voltammetry (ASV)

This is probably the most sensitive technique for lead, and in very low level work the avoidance of contamination is likely to be a more major problem than the approach of the analytical detection limit. Sensitivity may be improved even further by the use of alternative electrodes (e.g. glassy carbon) in place of the commonly used hanging mercury drop. Advantages of ASV are the possibility of simultaneous determination of four metals (Pb, Cd, Cu, Zn) in one sample, and the freedom from matrix effects as a standard additions technique is always used.

8.2.3 Colorimetric analysis

For many years the use of dithizone as a colorimetric reagent for lead was almost standard practice. Nowadays it is rather rarely used because of a range of disadvantages. These include the rather modest sensitivity and the possibility of interference from other metals [1]. In particular, the time-consuming nature of the method and the considerable experience necessary to achieve reliable results discourage many potential users. The method may still however be of use in laboratories lacking instrumentation for the other analytical methods.

8.2.4 Atomic absorption spectrometry (AAS)

As indicated above, this method is extremely popular because of its low cost and ease and rapidity of use. The simplest form of AAS is the flame technique which is of rather limited sensitivity. This may often prove suitable for analysis of lead in air and soils, but is insufficiently sensitive to analyse natural waters or most vegetation samples without a prior preconcentration. For these latter samples, or for short-term air samples flameless atomic absorption is required. Flame techniques are relatively free of interference and matrix effects, but in many applications the method of standard additions [1, 2] is necessary to minimize matrix interferences. Flameless AAS is more subject to interference by background absorption and matrix effects than the flame method, and the use of both deuterium background correction and the standard additions method is usually advisable.

8.3 Sampling and analysis of environmental media

The ranges of concentration which may be commonly encountered in environmental samples are summarized in Table 8.2. These data should assist in determining the necessary sample size, or the optimum analytical method.

8.3.1 Lead in air

These are two basic sampling techniques for particulate lead in air: the low volume and high volume methods [1]. Low volume sampling involves drawing air at 1 to 10 dm^3 min^{-1} through a small diameter (25 to 50 mm) filter. Sampling times may vary between 10 min and 7 days depending upon the degree of airborne contamination and the analytical sensitivity available. The relatively small amounts of lead which are collected in low volume sampling necessitate the use of a filter material with a low lead background. Most commonly used are cellulose ester membranes (Millipore, Sartorius, Gelman, etc.) of 0.22 or 0.45 μm pore size which are highly efficient collectors of particles of all sizes (including those smaller than the filter pore size) and have a low metal blank. Skogerboe *et al.* [3] have raised doubts as to the particle collection efficiencies of membrane filters at very low loadings, but their efficiency is generally believed to approach 100% in most applications. Leaky fitting in filter holders is often the cause of inefficiency when it does occur. Other filter types commonly used for particulate lead sampling are Whatman No 41, known to be markedly less efficient than membrane filters, glass fibre, efficient but with a rather high lead blank, and Nuclepore.

In high volume sampling air is drawn through large (20 cm × 26 cm) glass fibre filters at a rate of some 2000 dm^3 min^{-1}. Sampling times are rather prolonged (usually 24 h) to overcome the lead blank of the filter material. It is rare to analyse the whole of these filters for lead: usually a square (about 3 cm × 3 cm) or strip is cut from the filter for lead analysis.

Table 8.2 Typical concentrations of lead in environmental samples.

Environment	Lead concentration
Urban air (particulate Pb)	0.5–10 $\mu g\ m^{-3}$
Rural air (particulate Pb)	0.01–0.25 $\mu g\ m^{-3}$
Surface waters	<0.001–0.1 $mg\ dm^{-3}$
Treated industrial effluents	0.1–10 $mg\ dm^{-3}$
Tapwaters	0.005–0.3 $mg\ dm^{-3}$
Soils, natural	10–200 $mg\ kg^{-1}$
Soils, polluted, or in mineralized areas	100–10 000 $mg\ kg^{-1}$
Vegetation, background (dry weight)	0.1–5 $mg\ kg^{-1}$
Vegetation, polluted (dry weight)	1–1000 $mg\ kg^{-1}$
Small mammals (dry weight)	3–20 $mg\ kg^{-1}$
Tetraalkyllead in air	0.002–0.2 $\mu g\ m^{-3}$

When analysis is by XRF the filter may be used directly for the analysis, with standards being prepared by drying lead-containing solutions, or filtering suspensions of lead salts onto clean filter surfaces [1]. For AAS, ASV or colorimetric analysis, however, the lead must be leached into solution. Leaching agents vary from 2N HNO_3 with ultrasonic agitation or heating to the use of hot concentrated oxidizing acids (e.g. HNO_3, $HClO_4$, aqua regia). In general the less concentrated acids are now preferred, with high extraction efficiencies having been obtained for 3N HNO_3 used with either heat or ultrasonics [4]. The resultant solution, possibly after dilution with water, is used for the analysis. When hot concentrated acids are used, these are evaporated to dryness (or to low volume in the case of $HClO_4$) and the residue taken up in dilute acid for subsequent analysis.

Analysis of tetraalkyllead in air is more complex than determination of particulate lead. A prior separation of particulate and gaseous organic lead is usually achieved by filtration. Many reagents and techniques are then available for collection and analysis of tetraalkyllead, but a considerable number are unreliable as they are also sensitive to inorganic lead [5]. The recommended technique is that devised by Hancock and Slater [6] and modified by Birch *et al.* [7] for 24 or 48 h sampling. This method involves collection of tetraalkyllead in iodine monochloride solution, where it is converted into dialkyl lead ions. These are then selectively extracted by dithizone in carbon tetrachloride whilst inorganic lead remains complexed in the aqueous phase due to the addition of ethenediamine tetraacetic acid (EDTA). After back-extraction of the organic lead into dilute nitric acid/hydrogen peroxide, the levels are determined by flameless AAS.

8.3.2 Lead in water

Lead may exist in many forms in waters, both dissolved and suspended. As indicated in Chapter 3 there is no clear cut-off point between dissolved and suspended material and the arbitrary standard of a 0.45 μm membrane filter is usually adopted to separate 'filterable' and 'non-filterable' lead.

The non-filterable metal (that collected on the filter) may be analysed directly by XRF after drying or may be dissolved and analysed by an alternative technique. The methods of dissolution are usually rather different than for airborne particulates due to the presence of organic materials which require heating with strong oxidizing acids to destroy them and free the lead. Generally the filterable particles and filter are digested with hot concentrated HNO_3 or $HClO_4$ prior to taking up the lead in dilute acid for analysis.

Filterable metal may be analysed directly by AAS. Except in highly polluted waters, the use of flameless techniques will be necessary and research has shown that reliable results are obtained only by use of the method of standard additions to minimize matrix interferences. The sensitivity of flame methods may be enhanced if the lead is preconcentrated by extraction with ammonium pyrolidine dithiocarbamate in methyl isobutyl ketone (APDC/MIBK). This also eliminates many interferences, but prior heating of the sample with acid may be necessary to release complexed lead [8]. The use of ASV for analysis of filterable lead in water is rather more complex than might appear at first sight, since ASV is only sensitive to labile (weakly-bound) lead, and not to that strongly bound in complexes. Reduction of the sample pH by addition of an appropriate acid may cause release of most of the lead enabling analysis to be carried out. In general, AAS is to be preferred.

8.3.3 Lead in soils and sediments

The method of analysis selected will depend upon the degree of pollution of the soil or sediment. In unpolluted or scarcely polluted soils and sediments a substantial proportion of the lead is firmly bound in the silicate lattices of refractory minerals and can only be released by the use of hydrofluoric acid. Hence Harrison and Laxen [9] found HF/HNO_3 to be the most efficient of a range of reagents for extraction of lead from soils. In a heavily polluted soil, the major part of the lead is far more readily extracted and HNO_3 or HNO_3/HCl is of quite adequate efficiency. Soil is digested in the acid which is evaporated to dryness, causing oxidative decomposition of the organic materials present. The digest is then leached with dilute acid and filtered to provide a sample for analysis. Some sediments contain a very high proportion of organic material which may necessitate the use of $HClO_4$ for a complete oxidation. Due to risks of explosion the use of this reagent is recommended only after prior digestion with HNO_3 until all HNO_3-oxidizable material is destroyed, and evaporation of $HClO_4$ to dryness is not advised. If the final analysis of an acid extract is by AAS, the standard additions procedure is strongly recommended [9].

8.3.4 Biological materials

Biological samples, such as leaves, mice or fish require digestion with hot concentrated oxidizing acids (HNO_3, $HClO_4$) prior to analysis to destroy organic matrices. Precautions are necessary in the use of $HClO_4$ to avoid explosive hazards associated with this reagent (Section 8.3.3). After adequate destruction of the matrix, which may require several digestion procedures, the residue is leached with hot dilute acid, filtered and made up to volume for analysis by any appropriate method. A novel technique for the analysis of metals in sewage sludges which involves a minimum of sample pretreatment has been recently described [10].

8.4 Contamination during lead analysis

It will be appreciated from the earlier chapters that lead is present in many environmental media, and hence contamination during sampling or analysis is a real possibility. Almost all media yield some lead and the recommended materials for use in sampling and analysis are PTFE, borosilicate-glass (not soda-glass) and polyethylene (much preferred to polypropylene). Pre-cleaning of all apparatus, preferably by leaching with 10–50% HNO_3 for 24 h is essential, and prolonged exposure of any sample or apparatus to the atmosphere is to be avoided. If due care is exercised in all operations, very low levels of lead may be handled with confidence of obtaining a reliable analysis. When ultra-trace analysis is performed, i.e. involving concentrations $<0.1\ \mu g\ dm^{-3}$ very thorough precautions are necessary. Many workers have adopted the use of clean rooms, air conditioned with filtered air, for this type of work.

One very valuable check upon contamination is provided by running standard blank analyses. These provide information upon levels of lead in reagents and clean filters, as well as indicating contamination when it occurs. Blanks should be a normal part of every trace metal analysis.

References

[1] Harrison, R. M. (1977), Metal Analysis, in *Handbook of Air Pollution Analysis*, (ed. R. Perry and R. J. Young), Chapman and Hall, London.

[2] Beukelman, T. E. and Lord, S. S. (1960), The Standard Addition Technique in Flame Spectroscopy, *Appl. Spectros.*, **14**, 12–7.

[3] Skogerboe, R. K., Dick, D. L. and Lamothe, P. J. (1977), Evaluation of Filter Inefficiencies for Particulate Collection under Low Loading Conditions, *Atmos. Environ.*, **11**, 243–9.

[4] US Environmental Protection Agency (1978), National Ambient Air Quality Standard for Lead, *Federal Register*, **43**, 46246–77.

[5] Harrison, R. M. and Perry, R. (1977), The Analysis of Tetraalkyllead Compounds and their Significance as Urban Air Pollutants, *Atmos. Environ.*, **11**, 847–52.

[6] Hancock, S. and Slater, A. (1975), A Specific Method for the Determination of Trace Concentrations of Tetramethyl and Tetraethyllead Vapours in Air, *Analyst*, **100**, 422–9.

[7] Birch, J., Harrison, R. M. and Laxen, D. P. H. (1980), A Specific Method for 24–48 h Analysis of Tetraalkyllead in Air, *Sci. Tot. Environ.*, **14**, 31–42.

[8] HMSO (1976), Methods for the Examination of Waters and Associated Materials: Lead in Potable Waters by AAS, HMSO, London.

[9] Harrison, R. M. and Laxen, D. P. H. (1977), A Comparative Study of Methods for the Analysis of Total Lead in Soils, *Water Air Soil Pollut.*, 8, 387–92.

[10] Lester, J. N., Harrison, R. M. and Perry, R. (1977), Rapid Flameless Atomic Absorption Analysis of the Metallic Content of Sewage Sludges, I. Lead, Cadmium and Copper, *Sci. Tot. Environ.*, 8, 153–8.

Index